ZEW Economic Studies

Publication Series of the Centre for
European Economic Research (ZEW),
Mannheim, Germany

ZEW Economic Studies

P. Capros · P. Georgakopoulos
D. Van Regemorter · S. Proost
T.F.N. Schmidt · H. Koschel · K. Conrad
E.L. Vouyoukas

Climate Technology Strategies 2

The Macro-Economic Cost and Benefit
of Reducing Greenhouse Gas Emissions
in the European Union

With 28 Figures
and 55 Tables

Physica-Verlag

A Springer-Verlag Company

ZEW

Zentrum für Europäische
Wirtschaftsforschung GmbH

Centre for European
Economic Research

Series Editor
Prof. Dr. Wolfgang Franz

Authors

Prof. Pantelis Capros
Dr. Panagiotis Georgakopoulos
National Technical University of Athens
Department of Electrical &
Computer Engineering
42 Patission Street
10682 Athens, Greece

Denise Van Regemorter
Prof. Stef Proost
Center for Economic Studies
Catholic University of Leuven
69 Naamsestraat
3000 Leuven, Belgium

Dr. Tobias F. N. Schmidt
Henrike Koschel
Centre for European Economic
Research (ZEW), L7,1
68161 Mannheim, Germany

Prof. Dr. Klaus Conrad
Fakultät für Volkswirtschaftslehre
und Statistik
Universität Mannheim
Seminargebäude A 5
68161 Mannheim, Germany

E. Lakis Vouyoukas (Editorial Work)
Kodrou 13
10558 Athens, Greece

ISBN 3-7908-1230-7 Physica-Verlag Heidelberg New York

Cataloging-in-Publication Data applied for
Die Deutsche Bibliothek – CIP-Einheitsaufnahme
Climate technology strategies / ZEW, Centre for European Economic Research. P. Capros ... – Heidelberg;
New York: Physica-Verl.
   Vol. 2. The macro-economic cost and benefit of reducing greenhouse gas emissions
   in the European Union: with 55 tables. – 1999
   (ZEW economic studies; 4)
   ISBN 3-7908-1230-7

Cover design: Erich Dichiser, ZEW, Mannheim

SPIN 10733427   88/2202-5 4 3 2 1 0 – Printed on acid-free paper

Preface

This book provides a synthesis of the macro-economic results of the large-scale project entitled "Climate Technology Strategy within Competitive Energy Markets" partly financed by the European Commission under the Non Nuclear Energy (JOULE III) EU-RTD programme (Contract JOS3-CT95-0008). Research in this project involved several organisations from almost all European Union member-states, including: ICCS/NTUA (co-ordinator), CES/KULeuven, GRETA Associati, University of Mannheim/ZEW, ESD Ltd., IEPE/CNRS, IDEI, IPTS/JRC, UCL/CORE, SSE/EFI, KFA/STE, ENSPM-FI/CEG, University of Strathclyde, IER/University of Stuttgart, VTT, ETSU, IIASA/ECS, ERASME/CCIP, ECOSIM Cons. Ltd., OME, BPB, CRES.

In relation of the "pre-and post-Kyoto" context for Climate Change and also with respect to the preparation of the Vth EU-RTD Framework Programme, a large exploitation of the results is being made.

Huguette Laval of the European Commission, DGXII, was the project scientific officer. With Pierre Valette, Head of the Unit in DGXII, they have given fruitful advices for the project. Thanks are given to them.[1]

---

[1] The views expressed in this volume are those of the authors and do not reflect the views of the Commission or its officials.

Table of Contents

X

Executive Summary

There is a growing need at the European Union level for concrete decisions concerning the implementation of specific policies that aim at reducing atmospheric emissions, often termed QELROs (Quantified Emission Limitation and Reduction Objectives). Any policy mix that will be considered for implementation has to take into account a number of inter-related issues: inequalities in burden sharing and ways to alleviate them (for example, through income re-distribution or exemptions); utilisation of appropriate policy instruments (taxes, pollution permits, voluntary agreements); definition of accompanying measures that attempt to reduce the adverse economic implications of the policy so that environmental protection does not endanger economic growth and competitiveness.

This part of the research provides a systematic quantitative analysis of the economic cost of reaching alternative emission reduction targets for the EU and assesses alternative policy instruments in terms of their implications for the economy of individual EU member states. This analysis is the first systematic study of the macro-economic cost of emission abatement covering all EU member states (with the exception of Luxembourg) and several policy instruments. It is also the first time that a model utilises detailed observed Social Accounting Matrices for each country, including details on income re-distribution. The study also internalises the benefit from environmental protection.

Policy Evaluation Criteria and Main Results

The marginal abatement cost of emission reduction is generally significant. The level of the abatement cost varies depending on the type of accompanying policies adopted. The impacts on economic sectors, households and countries are significant.

Concerning the double dividend issue, environmental policy is likely to affect consumer's welfare negatively, but can have positive effects on employment and investment. Employment gains can be effected, by using the revenues from the carbon tax to subsidise labour costs. Investment can be encouraged when tax revenues are used to subsidise investment costs. Under extreme assumptions, the employment dividend disappears.

In most cases environmental benefits outweigh the costs (in terms of consumer's welfare) but total activity is generally reduced and competitiveness deteriorates, raising the issue of the sustainability of the gains.

The precise distributional effects also depend on the accompanying policies. In "labour recycling" cases, demand is re-oriented towards the consumer goods industries and away from energy intensive and capital goods industries. This structural shift has negative implications on investment and, probably, strong negative long-term implications.

Reducing the Cost of Abatement Policy

Reducing CO_2 emissions can invoke serious global changes in the economic environment of the European Union. Because of its all-pervasive character, CO_2 policy affects most economic sectors and agents. Limiting GHG emissions may incur considerable economic costs and have important burden sharing implications across countries, industrial sectors and economic agents. One possible way of reducing the costs of policy is to exploit existing distortions and deviations from optimal economic conditions. Three such deviations have been examined.

- *Distortionary taxation in the labour market and in investments:* The results indicate that it is indeed possible to get employment gains or investment increases through carbon taxes. However the impact on GDP is always negative and the impact on consumer's welfare, which crucially depends on labour market assumptions, also tends to be negative.

- *Terms-of-trade potential:* Alternative assumptions on the degree of exposure to foreign competition may result in limited such gains from terms of trade changes, showing that there may be a limited possibility for gains vis-à-vis the rest of the world.

- *Technological possibilities:* If there are energy-saving technology options that, while not significantly more expensive than existing ones, they are not adopted by firms and households then there are low-cost abatement possibilities. If such technologies are plentiful, then technology-driven sustainable development may be possible.

Burden Sharing

Allocating the burden of policy among country or among economic agents and sectors is the focus of interest in the international negotiation process on climate protection. As the review of equity rules exhibits, the range of principles and preferences that can be applied is wide. With respect to the self-interest of nations it is, therefore, no surprise that international agreements are difficult to attain. For a comparable homogenous group of countries like the EU, the choice of the operational rule is less decisive.

The results of an EU-wide tradable permit scheme indicate that, even within the EU, the burden sharing issue matters. The ability-to-pay rule, which favours the poorer and puts the richer countries at a disadvantage, implies higher overall welfare costs for the EU than the sovereignty rule, where permits are grandfathered with respect to a uniform reduction rate.

If big countries (i.e. countries that are powerful in terms of economic activity) are affected considerably, the interconnection of countries through bilateral trade might make the underlying burden sharing rule less attractive even for those countries which are favoured by the particular equity rule in terms of the initial allocation. This is the case for Austria, Finland, Ireland and Sweden under the ability-to-pay allocation. These countries could prefer an allocation according to the sovereignty rule, even though they could receive fewer permits in the initial allocation.

With respect to economic and environmental welfare, the sovereignty rule seems to be the most acceptable for an implementation of an EU-wide permit system. All countries show a positive welfare effect and the overall EU benefit is greater than under egalitarianism and ability-to-pay.

Derogations

Exemptions for the energy intensive sectors could lead to these industries suffering much less. It may even be possible that these sectors could benefit from lower labour costs, subsidised investments or from selling permits. However the additional burden facing the other sectors for any given CO_2 has negative implications for the economy as a whole. The overall conclusion is that the overall effectiveness of the policy is lower when energy intensive industries are exempted.

Derogations at the national level are not effective from an economic point of view and should only be selected on the basis of other considerations (e.g., cohesion targets) and then only if accompanied by some specific policy to make the period of exemption more fruitful. Under some circumstances, in countries with no environmental constraint lower prices will augment their market, but only for the period of the derogation. When these countries finally start imposing a constraint on emissions then they have to undergo a transitory period, which the other countries have now overcome. Thus, this policy is likely to lead to some short-term gains followed by longer-term losses. The overall impact is rather limited.

1 Introduction[2]

1.1 The Economic Cost of CO_2 Abatement

There is a growing need at the European Union level for concrete decisions concerning the implementation of specific policies that aim at reducing atmospheric emissions, often termed QELROs (Quantified Emission Limitation and Reduction Objectives). As the on-going debate points out, limiting GHG emissions may incur considerable economic costs and have important burden sharing implications across countries, industrial sectors, and economic agents.

Any policy mix that will be considered for implementation has to take into account a number of inter-related issues: inequalities in burden sharing and ways to alleviate them (for example, through income re-distribution or exemptions); utilisation of appropriate policy instruments (taxes, pollution permits, voluntary agreements); definition of accompanying measures that attempt to reduce the adverse economic implications of the policy so that environmental protection does not endanger economic growth and competitiveness.

This volume presents a systematic quantitative analysis of the economic cost of reaching alternative emission reduction targets for the EU and assesses alternative policy instruments in terms of their implications for the economy of individual EU member states. All simulations presented have been conducted with the *GEM-E3* computable general equilibrium model[3].

---

[2] All research in this volume was conducted within the "Climate Technology Strategy within Competitive Energy Markets", JOULE project of DGXII.

[3] The GEM-E3 model was built under the auspices of DGXII by a consortium involving among others, NTUA, KUL, ZEW, IDEI and University of Strathclyde. The model has been extensively used by several directorates of the European Commission, especially on the issues of general tax reform, double dividend and (in a version modified by NTUA and KUL) for the evaluation of the effects of the Single Market Programme.

In more concrete terms, the volume attempts to tackle a number of specific questions.

- What is the cost of reducing CO_2 emissions in the European Union? What is the impact on consumer's welfare, on GDP, on competitiveness and on employment?

- What is the shape of the abatement cost curve for the European Union? Is there a potential for relatively "cheap" abatement (known as "no-regrets" potential)?

- How do different policy instruments for the environment (pollution permits or taxes) affect the economy?

- How can the environmental burden be shared among countries and what are the implications for each Member State and economic agent (firms, households)?

- Is there a potential for economic dividends as a "by-product" of the re-distribution of income effected by the environmental policy? What are the accompanying measures that need to be considered and how do they affect the economy?

- Internalising environmental externalities tackles the whole problem of environmental pollution (as opposed to global warming). Could such a policy be more efficient in terms of the environment and if so, what are its economic implications?

Compared to the literature the simulations with the GEM-E3 model presented in this volume constitute an improvement with respect to the state-of-the-art in a number of ways.

This is the first systematic study of the macro-economic cost of emission abatement covering all EU member states (with the exception of Luxembourg) and several policy instruments. It is also the first time that a model utilises detailed observed Social Accounting Matrices for each country, including details on income re-distribution. The study also internalises the benefit from environmental protection, as this was quantified through the EXTERNE data.

Despite its attractive features a number of shortcomings of the present research should also be noted: the base year data of the model is somewhat outdated [4]; the model is not currently linked to an engineering energy model; technological progress is largely exogenous and only a first attempt at its endogeny is presented in some simulations; the representation of the labour market is rather simplified

---

[4] The only year for which EUROSTAT provides input-output tables for all EU countries is 1985.

and does not include a level of disaggregation sufficient to allow for burden sharing analysis among social groups[5].

1.2 Structure of the Volume

The structure of the volume is as follows.

Chapter 2 introduces the main questions as these have been addressed in the literature. The distinction between bottom-up and top-down approaches is clarified and the most well known models and methodologies are presented. This chapter also hints at the relative advantages of the various types of models and their relevance for policy analysis.

Chapter 3 presents the policy instruments that can be used to curb GHG emissions, highlighting those that can be simulated with the GEM-E3 model. It then provides a systematic set of criteria for their assessment.

Chapter 4 gives an overview of the simulations conducted with the GEM-E3 model. It presents the baseline scenario for the European Union member states, which is the reference for all subsequent model applications. An overview of the mechanisms through which environmental constraints affect the model concludes the chapter.

Chapters 5 and 6 provide the main set of simulations. The cost of reaching a wide range of emission reduction targets (-5 to -25% from the baseline, or equivalently +3% to -17% from 1990) is computed under the assumption of a pollution permits market, or tax. In the latter case two types of accompanying measures are examined (reductions of labour costs or investment subsidisation).

Chapter 7 is devoted to a first examination of the role technological progress in energy as an endogenous decision of economic agents. The simulations presented employ a modified version of GEM-E3, which incorporates a depletable energy-saving potential that can be acquired by the economic agents.

Chapter 8 discusses the issue of the internalisation of external (in this case environmental) costs. Such a policy aims at applying this principle to the externalities generated by energy consumption. The goal is to evaluate the impact of an integrated policy approach towards air pollution compared to a policy addressing only the global warming issue (as in previous chapters).

Chapter 9 examines the burden sharing issue. All analysis of emission reduction in chapters 5 to 8 did not pay much attention to the distribution of efforts among countries: the effort was allocated so that the total EU-wide cost was minimised.

---

[5] All the above issues are currently under development as part of a new JOULE programme.

However equity or political criteria may imply that a different allocation is desired. Several alternative assumptions on the allocation of effort among countries are examined in this chapter.

Chapter 10 performs sensitivity analysis in two main streams. The first concerns the sensitivity of the results to the modelling assumptions adopted in GEM-E3 (such as the behaviour of the labour market or the reaction of the rest of the world). The second examines the effectiveness of exempting some or all industrial sectors from the impact of environmental policy and of postponing action in the cohesion countries for a few years.

Chapter 11 summarises the main findings, the shortcomings of the research and highlights the issues that are left unanswered by this volume.

2 Literature Review

2.1 Introduction

The threat of climate change, possibly generated by the growing accumulation of carbon dioxide in the atmosphere, has become a major economic and political issue. As a consequence, an increasing number of empirical economic models has been developed in recent years to address questions related to the economic effects on the economy of policies designed to control the global environment. In particular these models have attempted to provide a quantitative analysis of future trends in energy consumption and other economic variables, and to assess the economic costs of reducing carbon and other emissions (accurate surveys of these models are contained in Grubb et al., 1993a; Boero et al., 1991).

Although most of the available literature on the costs of GHG (green house gases) mitigation has been written in the period since 1988, interest in this issue began more than a decade earlier with Nordhaus (1977; 1979) whose contribution was followed by around a dozen pioneering studies. The picture has evolved since then, but it retains two features from this earlier period.

- The focus is on CO_2 emissions and the other GHGs are either ignored or treated separately in an ad hoc manner. This is not only due to the importance of CO_2 emissions relative to those of other GHGs, but also due to the existence of an established tradition in long-term energy modelling and forecasting. Because of the long lead times required for planning of energy supply systems, and the instability of energy markets by the mid-1970s, analysts began to develop long term energy demand and supply models. These models could be used to analyse the CO_2 emissions associated with the energy system simply by adding emission coefficients to the model. Such a modelling capability was not available with respect to other GHGs.

- The structure of the debate was determined very early by the wide range of numerical results about the potential for, and the costs of, mitigation: some studies argued that mitigation policies were likely to entail very substantial costs (Nordhaus, 1977), while others concluded that the costs would be relatively small (Edmonds and Reilly, 1983) and still others that these costs might in fact be negative and bring an overall benefit (Lovins et al., 1981).

This debate became more heated, and the demand for this type of study accelerated, after the drafting of the UN Framework Convention on Climate Change in 1992. The controversies among modellers about the overall costs of GHGs mitigation strategies made it clear that statesmen, policy makers, business people and journalists are sceptical of figures coming from economic modelling exercises. Part of the scepticism stems from the past failures of predictions of future energy demand; another part, however, stems from misunderstandings about what economic and energy models are able to capture and about how to use their results. A final source of potential problems is that many agencies have employed models that were not initially designed to highlight the cost of emission reductions.

In any economic analysis, the cost of mitigation is calculated as a difference in costs between a reference situation and a new one characterised by lower emissions. The analysis usually computes 4 types of costs:

i.   the direct engineering and financial costs of specific technical measures;

ii. the economic costs at the level of a sector or economic agent;

iii. the macro-economic costs;

iv. the welfare costs.

Given the global dimension of the environmental policy and the synergies between policy measures, a number of other effects need also to be taken into account:

i.   the negative cost potential, namely mitigation caused by more efficient technologies whose costs are lower than the technologies currently in use;

ii. potential economic dividends, such as the possible positive effects on growth or employment of the recycling of carbon tax revenues or of the technological externalities associated with fostering research and development programs;

iii. multiple environmental dividends, caused by the synergy between GHG mitigation strategies and the mitigation of other environmental problems such as local air pollution, urban congestion or land and natural resource degradation.

The existence of these positive side effects within a suite of mitigation measures would result in lowering the gross cost of these measures.

2.2 Models and Studies for the Cost of GHG Abatement

2.2.1 Top-down and Bottom-up Models

Historically, the debate in the energy field has been framed by the distinction between top-down and bottom-up methodologies. Both classes of models have been used to answer the question of how much it would cost to limit GHGs emissions. The results from each type of model naturally reflect the choice of the modelling methodology. The differences between models reside in emphasis: while a degree of generality is desirable in modelling, no model can cope equally well with all dimensions of the GHG abatement issues.

- *Top-down models* are based on the analysis of the micro-economic behaviour of the economic agents (firms, households, and government) and their interactions, taken at some level of aggregation. Environmental policy affects the allocation of resources in the economy through changes in relative prices; Top-down models provide a consistent framework for analysing the economic implications of policy measures, but are usually criticised as lacking engineering evidence for the energy system;

- *Bottom-up models* are much more accurate in the representation of the energy system's technical details on energy-technology choices and costs. These models can evaluate more accurately the direct costs of environmental policy, but are usually weak in determining interactions and feedback of such changes with the rest of the economy.

The main findings from both types of models are presented in the following paragraphs. In most cases, the two approaches yield different outcomes: the bottom-up models tend to suggest that the costs of GHG mitigation might be low. The reason for that is the (implicit or explicit) assumption that there exists a wide range of technological opportunities that the economy will be able to exploit immediately and at low cost. Top-down models on the other hand, tend to suggest the opposite: the top-down approach assumes a world in which there is usually limited scope for technological improvement and cost effective opportunities. Thus environmental policy is usually found to affect the economy in a negative way. This persistent controversy has been present in the literature, portrayed as opposing the "optimism of the engineering paradigm to the pessimism of the economic paradigm" (Grubb et al., 1993a).

Bottom-up Energy Models

Bottom-up models aim at identifying alternative ways to provide energy services. The objective is pursued through a high level of disaggregation of the energy system (usually of engineering character), both on the supply and the demand side. The main limit of this approach is that they generally neglect feedback on the economy and rebound effects through international energy markets. Moreover, the process of innovation and the diffusion of new technologies are usually not considered. The *MARKAL* (see Rowe and Hill, 1989) and *EFOM* (of the European Community) energy models are some of the well-known bottom-up models developed for European countries.

Bottom-up models can be distinguished in a number of categories: the *simulation models* solve a simultaneous set of equations describing the way a given set of technologies is allocated. These models tend to use reduced form, econometrically estimated equations and provide a smooth projection for the short-medium term (see for example the *MIDAS* model). The *optimisation models* represent the energy system as a global optimisation problem (usually of the linear programming type). In the near past a third class of energy models, the *market equilibrium models* have emerged, with two main proponents: the *NEMS* model in the US, and the *PRIMES* model in Europe. The main characteristic of these models is that each agent is assumed to optimise her own behaviour. The agents then interact via demand and supply of energy goods. The prices and quantities of equilibrium are then derived as the result of the market clearing mechanism. The NEMS model even incorporates feedback from the energy system to the economy (which is exogenous in most other bottom-up models).

The major bottom-up studies for European Union countries have been: the COHERENCE study carried out for the CEC Joule Programme (1991); the studies for Denmark, France and Netherlands within the UNEP work (1994); the IPSEP study covering five countries (1993); the ETSAP study (Kram, 1993); the Swedish study by Bodlund et al. (1989). The primary aims of the European studies have been to examine the existence of energy savings potential with negative or low economic cost and to analyse the cost-effectiveness of energy supply and demand options for achieving a certain emission reduction goal. The study results are highly dependent on the assumptions on energy efficiency measures included in the baseline scenario.

Bottom-up research in the US and Canada has tended to suggest, as elsewhere in the world, that significant decrease in CO_2 emissions are possible without great cost to the economy. Compared with the top-down studies, the structure of formal models in bottom-up analysis is generally less important for the results. Instead, input assumptions are dominant and harmonized among models.

Bottom-up methods offer some advantages for dealing with transitional economies in that they focus on the physical stock of equipment and apply scenarios for its evolution, with less concern for anticipating macro-economic equilibrium conditions. However, the data on current equipment end-use energy efficiencies is severely limited. A comparison of emissions reduction studies reveals striking differences with the results of some top-down models. On the bottom-up side, a number of studies have been completed after the political reforms of the late 80s and early 90s. Some of this work was conducted using the EPA End-Use Energy Model.

Most bottom-up studies of developing countries focus at the national level. This contrasts with top-down studies, which tend to treat developing countries in groups (with the exception of China). Two major multi-country studies have been conducted by the Lawrence Berkeley Laboratory (LBL) and by the UNEP Collaborating Centre on Energy and Environment (UCCEE). These two centres facilitate comparative analyses in a similar way to the studies carried out by the EMF in the US. Only some studies have estimated the costs of achieving emission reduction: the UNEP studies (1994), the LBL study for India (Mongia et al., 1991) and the ADB study for China (Asian Development Bank, 1993).

Several bottom-up studies have been carried out with the aim of providing a normative view of the very long-term future. A typical global bottom-up analysis has been carried out for Greenpeace International by the Stockholm Environment Institute, Boston (1993) for the period 1985-2100. Three scenarios were compared: the FFES (fossil-free energy future) scenario, the IPCC 1991 scenario (high projection case of CO_2 emissions) and the EPA's Rapid Reductions (low projection case of CO_2 emissions). FFES expects solar/wind resources to cover as much as 79 percent of primary energy consumption in 2100 compared with only 9 percent in the EPA scenario. This high contribution of renewables contrasts to the conclusion of the extensive analysis of the renewable energy potential done by the World Energy Council, 1994. An integrated global top-down/bottom-up analysis has been carried out for the IPCC WG II Second Assessment Report (1994) resulting in the construction of scenarios for a Low Emissions Supply System (LESS).

Top-down Models

In the top-down approach, the energy-economy interactions are modelled at the macro level, for a single country, for a group of nations, or for the world as a whole. These models aim at evaluating the economic impact of measures to reduce greenhouse gas emissions. They allow for varying degrees of disaggregation, particularly of the energy sector. The top-down approach assumes a world in which there is usually limited scope for technological improvement and cost-effective opportunities. Hence, substantive action to curb GHGs is only achievable through costly changes. There are many types of models following this approach. In particular, three major ones may be distinguished: the traditional macroeconomic models, the technical-economic sectoral models, and the computable general equilibrium models (Carraro et al., 1994).

The traditional macro-econometric models (for example, HERMES, Capros and Karadeloglou, 1992; IEA medium term model, Vouyoukas, 1993) follow a neo-Keynesian theoretical approach. The assumed structure is *demand-driven* and under-utilisation of productive capacity is possible in the short and medium term.

They use lag equations to model *inertia* in the adjustment processes and allow for unemployment in the short run in response to shocks (some of them allow for structural unemployment due to inadequate demand for labour also in the long run).

A second type of top-down modelling approach is the technical-economic models, which generally combine a simplified description of the economic relationships with a detailed representation of the energy sector. The most important global models in this group are ERM (Barns-Edmonds-Reilly, 1992) and Global 2100 (Manne and Richels, 1992). They present different levels of detail of energy-supply technologies and contain a simplified representation of macroeconomic and trade linkages. They extend the analysis over a very long time horizon. Technical progress is represented by an exogenous index measuring the autonomous energy-intensity improvement (AEEI). Uncertainty is examined only in the work by Manne and Richels (1992).

For both these two classes, the fact that they assume a non perfect equilibrium economy explains why there is a gap between their findings in the case of non recycling or lump-sum recycling of tax revenues (which show high costs) and their findings in the case of efficient recycling (which often show an overall economic benefit from a carbon tax). For longer time periods, they fail to account for the effects of intertemporal preferences and expectations, and capture technical change in a rather static fashion.

The third type of top-down models belongs to the Computable General Equilibrium (CGE) class. These models focus on a medium/long term analysis of the effect of climate policies in the period after adjustment of the economy to short-term effects. They rely on a resource allocation principle and a market clearing mechanism for all goods which determines prices. They have in common the theoretical assumption of a Walrasian representation of the economy, but they differ with respect to many other aspects: the behaviour of economic agents and the market regimes; the calibration or the estimation of the model parameters, the time horizon, the geographic extension, the representation of technology options[6]. The fact that they are benchmarked on a given year, in order to guarantee the consistency of the parameters, allows a greater flexibility for using information coming from other models or expert judgements about possible shifts in current trends. The GEM-E3 belongs to this category, representing however only the EU countries. Among the global models, the Whalley and Wigle (1992) and the OECD's GREEN (Oliveira-Martins et al., 1992) are the best known.

Given an exogenous perturbation, CGE models produce a price-dependent general equilibrium response. However, because of their formal structure, they do not

---

[6] A detailed presentation of the general equilibrium methodology and the classification of economic models is provided in the first annex.

provide an accurate picture of the time path toward the equilibrium (they tend to understate transition costs).

Model Comparison Studies

One of the most well known study in the US was conducted by the Energy Modelling Forum of Stanford University (EMF, 1993). Fourteen top-down models employing common assumptions were used to analyse a standardised set of emission reduction scenarios. Although the focus was primarily on the US, many of the insights are applicable to developed countries in general.

In selecting parameters for standardisation, the EMF study focused on the most influential determinants of emission reduction costs: GDP, population, fossil-fuel resource base, cost and availability of long-term supply options. Different emission scenarios and taxes based on the carbon content of the fossil fuels were adopted. Two parameters have been recognised to be particularly important in explaining the differences in tax projections: the price elasticity of energy demand and the speed with which the capital stock adjusts to higher energy prices. The results of the adoption of the carbon tax showed a marked variation in GDP losses across models. The relationships among timing, emission reduction costs and R&D were also considered. Alternative methods of recycling the tax revenues (reducing budget deficit; reducing marginal rates of income, payroll, corporate or other taxes; granting tax incentives to preferred activities; increasing the level of government expenditures) have been pointed out in Shackleton et al. (1993).

One notable attempt at a systematic model comparison of non US OECD models was conducted by the OECD in the early 1990s (Dean and Hoeller, 1992). The exercise was patterned after the parallel study being undertaken by the EMF. The two studies used many of the same models.

Modelling emission reduction costs for the formerly planned economies has been and is still particularly challenging, and the problems of adapting top-down models explain the dearth of applications for this region of the world. In fact, his long history of highly subsidised energy prices and other inefficiencies in the structure of incentives are key issues in extending top-down models to emerging economies. The treatment of energy subsidies was faced by Dean and Hoeller (1992); the market distortions were underscored by Manne and Schrattenholzer (1993) using Global 2100 model and Manne and Oliveira-Martins (1994) using GREEN and 12RT models.

The analysis for developing nations is usually piggybacked on the developed country studies. Many assumptions of top-down models, such as future markets, perfect information, competitive economic dynamics on both demand and supply sides, optimising behaviour of producers, consumers and government, are often found to be invalid in developing countries. In addition, top-down models tend to underestimate the contributions from the informal sectors. Despite these limitations, they can provide insights into some issues like taxes and subsidies, revenue recycling, international trade, allocation for R&D and backstop technologies. In a recent study, OECD (1994) used GREEN to analyse the impact of subsidy removal in China and India.

The Impact of Revenue Recycling on GDP and Employment

It has been suggested that carbon or energy taxes could be used to reduce non wage labour costs and consequently, increase employment in Europe. European governments are increasingly concerned about the adverse social and budgetary consequences of high unemployment. A new European plan to reduce unemployment (Drèze et al., 1993), also known as Delors' plan, contains the suggestion that moving the tax burden away from employment towards undesirable pollution would provide a welcome boost to employment tax revenues. Shifting the tax system away from labour towards polluting sources would encourage employers to substitute labour for capital and other inputs, thereby making production more labour intensive at the aggregate level. If, at the same time, the impact on labour supply were negligible, this policy would reduce unemployment.

There are four issues that deserve careful discussion.

1. The *functioning of the labour market*. In order to assess the effectiveness of a gross wage reduction, it is necessary to understand the role played by unions in European labour markets (Layard et al., 1991; Holmlund and Zetterberg, 1991). Indeed, if a reduction of labour taxes (e.g. payroll taxes) is not reflected in gross wages, e.g. because unions bargain on the ex-post net wage, it is unlikely to observe an increase of employment. The GEM-E3 model, in several studies[7] has shown that the possibility of increased employment indeed exists, at least in the medium term. Other studies[8] argue that although such gains can indeed be temporarily effected, in the longer term wages tend to revert back to their baseline. As a consequence, the initial increase of employment tends to disappear in the long run. This fact can be rationalised in terms of the modelling of the labour market: wage bargaining implies that only in the short run unions are unable to offset the payroll tax reduction through an increase of net wages. In the long run, the fiscal change is completely absorbed by the change of net wages. Hence gross wages, other things being equal, go back to the initial level[9]. As a consequence, employment can increase only in the short run.

---

[7] Capros et al. (1996, 1997).

[8] Brunello (1996) and in Carraro et al. (1996).

[9] They further increase with respect to the baseline scenario because of two feedback mechanisms:

2. *Technology*. The effectiveness of the Delors' plan is based on the assumption that an increase of energy prices and a reduction of labour costs shifts factor demand from energy to labour. In other terms, labour and energy are assumed to be substitutes. However, the increase in energy prices is also likely to induce firms to introduce new environment-friendly technologies (Carraro and Topa, 1995, Capros and Georgakopoulos 1997), whose labour requirements may be lower than those of the old technology. Therefore, the technology shift may reduce the potential for employment gains. Recent analyses have shown that environmental taxes or emission permits are likely to be sub-optimal instruments to achieve the adequate level of technological innovation. For example, Laffont and Tirole (1994) show that, in a dynamic framework, taxes or permits can lead firms to over-invest in environmental innovation, in order to bypass the fiscal burden, thus reducing social welfare. On the other hand, in an endogenous growth model, Musu (1994) shows that emission charges alone are unable to induce firms to account for the positive R&D externalities that would lead to the appropriate investment in innovation; hence, firms tend to under-invest. A similar conclusion is attained in Carraro and Topa (1995), using an industrial organisation model of dynamic innovation. The paper shows that firms tend to delay innovation, rather than to under-invest, when the only policy instrument is environmental/energy taxation. Available empirical evidence suggests that energy efficient and/or environmental friendly innovations may also be labour saving, particularly in the electricity and energy sectors (Boetti and Botteon, 1994). Therefore, an environmental fiscal reform, by inducing technical progress may not favour an increase of employment.

3. *Tax distortions*. An environmental fiscal reform would generally be introduced into an existing fiscal structure, which is obviously a second best one (chiefly, because it is based on distribution goals). Therefore, it is by no means clear whether the tax shift will reduce or increase distortions. An increase in distortions, by reducing the efficiency of the economic system, may result in a reduction of employment. This issue has been examined by Bovenberg and De Moij (1994) and Parry (1994). They find that the fiscal reform in which environmental taxes replace some of the labour tax leads to no reduction (and usually an increase) in labour market distortions. Second, the tax on the environmentally damaging commodity induces changes in the commodity

---

- the unemployment reduction achieved in the short-run increases unions' bargaining power thus inducing, through the dynamics of the bargaining equation, a counter-effect in the long-run;
- the energy-saving R&D activity induced by the energy tax increases the growth rate of technical progress, thus increasing factor productivity and, as a consequence, firms' profits. The sharing rule, which is implicit in the wage bargaining equation, implies that gross wages are proportional to profits. Hence, in the long-run gross wages are larger than in the baseline.

market, which distorts choices among alternative commodities. These two distortionary effects, in labour and commodity markets, imply that, apart from environmental considerations, the revenue neutral combination of an environmental tax and reduction in labour tax involves a reduction of the environmental component of welfare. In fact the distortions in the commodity and labour markets are connected. To the extent that the environmental tax leads households to substitute other commodities for the tax commodity, there is a reduction in the gross revenue yield of the tax. This tax base erosion effect limits the extent to which the environmental tax can finance a reduction in the labour tax, and augments the overall gross costs of the fiscal reform. These results have been generalised by Bovenberg and Van der Ploeg in several articles. In Bovenberg and Van der Ploeg (1992), the authors investigate the issue of tax recycling to increase employment (higher emission charges, lower payroll taxes) in a second-best framework. The model considers a closed economy and assumes that emissions are a by-product of consumption activities. Moreover, all markets are assumed to be perfectly competitive and to clear. The conclusion is that a more ambitious environmental policy (higher taxes) typically reduces employment. In Bovenberg and Van der Ploeg (1993a) the previous model is generalised to consider an open economy in which pollution is a by-product of production, rather than of consumption activities. Moreover, the paper allows for substitution in production between labour, capital and natural resources. However, despite the factor substitution induced by a lower tax on labour and a higher tax on natural resources, the paper finds that higher environmental concern again reduces employment. It could be argued that this "negative" result originates from the assumption of a competitive and perfectly clearing labour market. This is why in two subsequent papers Bovenberg and Van der Ploeg (1993b, 1994) introduce wage rigidities into the model, thus allowing for unemployment in the labour market. The results are not more encouraging. As emphasised in Bovenberg (1994), several restrictive conditions must be met for an environmental tax reform to benefit employment by reducing the tax burden on labour: "First, the overall tax level should be contained. Second, the distribution of the tax burden should be moved away from workers to others, i.e. capital owners, the owners of resources, and the recipients of income transfers[10]". This implies that labour must be a better substitute for resources than the fixed factor (i.e. the capital stock), and that the fixed factor must play an important role in production. As stressed in Bovenberg and Van der Ploeg (1994), "this is unlikely to be the case in the long run and in open economies, especially as economic integration proceeds[11]".

---

[10] Bovenberg (1994).

[11] Bovenberg and Van der Ploeg (1994).

4. The type of *institutional framework,* which is the most appropriate to introduce an emission tax and the revenue recycling to boost employment. Should the environmental fiscal reform be designed at the central level by the European Commission, or should such a reform be decentralised according to the subsidiarily principle? Is the reform more effective if harmonized among different EU governments or can it be applied even unilaterally? This issue is addressed by Carraro and Galeotti (1996). The main conclusion is that changes of the institutional setting do not modify the qualitative features, which define the effects of the fiscal reform and only slightly modify its quantitative effects. However, if the main goal is employment relief, rather than emission reduction, the results suggest that a federal policy in which both tax rates and wage subsidies are harmonized is likely to be the best institutional setting. Tax rates have to be harmonized because decentralisation would lead to lower rates, which provide fewer resources to be recycled to boost employment. Wage subsidies have to be co-ordinated because revenue recycling produces the largest increase of EU employment when the tax revenue is used chiefly in those countries with high unemployment levels;

The small effects of environmental fiscal reform suggest that restricting the reform to emission charges and wage subsidies is probably inefficient. The idea of reforming the fiscal system by correcting externalities and by lowering distortionary taxes should be considered in more general terms, by proposing an overall change of the tax system. A comprehensive, well-designed reform is likely to provide larger gains in all countries.

European concern for the unemployment is not shared equally in North America, where quantitative analyses focused instead on the possibility that an environmental fiscal reform might reduce emissions and increase total welfare as well (the so called "strong" double dividend).

The models used to examine this issue in the US include the Goulder and Jorgenson-Wilcoxen intertemporal general equilibrium models of the US, the DRI and LINK econometric macroeconomic models of the US, and the Shah-Larsen partial equilibrium model[12].

In most cases, the revenue-neutral green tax swap involves a reduction in welfare, that is, entails positive gross costs. Results from the Jorgenson-Wilcoxen model, however, support the strong double dividend notion. A thorough examination of the differences in model outcomes, and an extensive test of how these differences account for different outcomes, is beyond the scope of this review. However, one potential explanation lies in the differences between the Jorgenson-Wilcoxen and

---

[12] The Shah-Larsen model is by far the simplest of the five models, in part because it takes pre-tax factor pricesas given. Despite its simplicity, the model addresses interactions between commodity and factor markets and thus incorporates some of the major efficiency connections.

Goulder models in the marginal excess burden (MEB) of capital taxation[13]. The interest elasticity of saving is higher in the Jorgenson-Wilcoxen model than in the Goulder model. In addition, the Jorgenson-Wilcoxen model assumes that capital is fully mobile across sectors, while the Goulder model includes adjustment costs, which limit the speed at which capital can be reallocated and lower the elasticity of capital demand. Thus, elasticities of capital supply and capital demand are higher in the Jorgenson-Wilcoxen model; correspondingly, the marginal excess burden of capital taxation is considerably higher in the Jorgenson-Wilcoxen model than in the Goulder model, and the difference in the marginal excess burdens of capital and labour is larger. In the Goulder model, the difference in MEBs per dollar is $0.1; while in the Jorgenson-Wilcoxen model appears to be considerably higher. A large deviation in the MEBs on capital and labour taxes works in favour of the strong double dividend (particularly if the burden of the environmental tax falls on labour). This helps explain why, in the Jorgenson-Wilcoxen model, a revenue-neutral combination of carbon tax and reduction in capital tax involves negative gross costs (that is, a positive change in gross welfare). It is more difficult to account for the fact that substituting a carbon tax for a labour tax involves negative gross costs in the model.

Like the theoretical results, the numerical outcomes of the above studies tend to weigh against the strong double dividend claim. But there is less than perfect agreement among the numerical results. Discerning the sources of differences in results across models is difficult and frustrating, in large part because of the lack of relevant information on simulation outcomes and parameters. Relatively few studies have performed the type of analysis that exposes the channels underlying the overall impacts. There is a need for more systematic sensitivity analysis, and for a closer investigation of how structural aspects of tax policies (type of tax base, narrowness of tax base, uniformity of tax rates, etc.) influence the outcomes. The present volume attempts such a detailed analysis of the conditions under which economic dividends can be expected from environmental policy.

International Co-operation, Emission Rights and Carbon Leakage

A growing number of studies attempt to provide a global perspective of the assessment of abatement costs. Significant reductions in emissions can be accomplished only through international accords and co-operation. Moreover, actions taken in one region tend to have *spillover* effects into other regions. Partial equilibrium analyses ignore significant linkages that could substantially alter the economic impacts of carbon constraints.

Emission reduction actions affect global emissions and atmospheric concentrations, irrespective by their geographic and sectoral origins. The least-

---

[13] Of the five models, these two are the most similar and allow for the most straightforward comparisons.

cost global abatement strategy requires reducing emissions where it is cheapest to do so. To the extent that marginal costs vary among regions, there are opportunities for efficiency gains through international co-operation. These opportunities may be exploited through a system of international trade in carbon emission rights or through a global carbon tax.

Emission rights trading would allow for a more efficient allocation of emission reductions across regions by letting countries trade to the point where the marginal cost is the same across all and activities. A global carbon tax would result in the marginal cost of emission reduction being equal for all countries. The carbon tax that would be required to achieve a given level of emission reduction fixed on a region-by-region basis (OECD, 1994) varies widely among regions. Efficiency gains can be obtained by shifting abatement from high to low marginal cost regions: if the resulting cost savings are shared between participating regions, all can be made better off.

The establishment of international trade in carbon emission rights requires a decision on the allocation of emission rights among nations. This has major implications for the international distribution of wealth. Edmonds et al. (1993) studied a variety of schemes for allocating emission rights. One extreme option they considered, was a *grandfathered* emissions principle in which future emission rights are allocated on the basis of the share in global emissions at the time of joining a global agreement; In another extreme scheme, they examined an *equal per capita emissions* principle where emission rights are allocated on the basis of regions' shares in adult population. The allocation of emission rights differs markedly under these two extreme quota allocation schemes: OECD countries hold on to approximately fifty percent of the emission rights when the status quo is maintained under a grandfathering scheme, while under the equal per capita emissions scheme their share drops to about twenty percent. A third scenario is referred to as a *no harm to developing nation* principle. Here developing nations receive sufficient emission rights to cover their own emissions and to generate sufficient revenue from excess quota sales to cover the economic cost of participating in the agreement.

Some proposals for limiting CO_2 emissions call for high-income countries to take the lead in reducing emissions. When abatement actions are limited to a subset of regions, it is important to consider the so called *carbon leakage* effects, which represent the impacts of the emission policies of abating regions upon the emission levels of non abating regions. Such leakage effects can be positive or negative. There have been several attempts to estimate leakage (Barrett, 1994; Pezzey, 1992, by using the model of Walley and Wigle; Horton et al., 1992; Oliveira-Martins et al., 1992, with the *GREEN* model; Burniaux and Oliveira-Martins, 1993; Manne and Oliveira-Martins, 1994).

Among the most recent studies on international environmental agreements (IEA), Barrett (1997) and Carraro (1997) present some theoretic results stemming from a game-theory perspective of the structure of these agreements. According to Karp (1997), a successful IEA depends on the characteristics of pollution (local or global), on its duration, and on agents' planning horizon. Moreover, the patterns of negotiation and the importance of the revelation of the self-interests of countries are dealt with in Hourcade (1997).

Applied studies that are supported by modelling include Nordhaus and Yang (1996), Yang (1997), Bollen and Gielen (1997) and Hinchy and Fisher (1997). Nordhaus and Yang analyse different national strategies in climate change policy (pure market solutions, efficient co-operative outcomes, and non co-operative equilibria) with the *RICE* model. Their findings are that i) co-operative policies show much higher levels of emissions reductions than non co-operative strategies ii) there are substantial differences in the levels of controls in both co-operative and non co-operative policies among different countries and iii) high-income countries may be the major losers from co-operation.

Yang explored the conditions necessary for successful stabilisation agreements by simulating the OECD's model GREEN. He considered the trajectories suggested by IPCC and Wigley, Richels and Edmonds (both aiming at a concentration of 550 ppmv) and concluded that i) this path will be very difficult under any circumstances ii) inclusion of FSU and EE in the developed countries group is problematic iii) there is great uncertainty in forecasting of relative wealth and iv) any stabilisation agreement will require continued renegotiations.

Bollen and Gielen analysed the issue of tradability of regional emission permits by comparing a globally co-operative emission reduction policy with a multilateral domestic policies in all regions without tradability using the *WORLDSCAN* dynamic applied general equilibrium model for the world economy.

Finally, the general equilibrium model *MEGABARE* has been used by Hinchy and Fisher to examine the concept of costs of abatement as costs of production of a global public good. Such an approach has the appeal of simultaneously dealing with both equity and efficiency issues. But general problems in negotiating an agreement under such an approach are still present.

Other Greenhouse Policy Instruments

The variety of instruments available to policy makers include regulatory instruments (uniform technology and performance standards, government investment, voluntary agreements, and other non market based instruments), market-based instruments (taxes, subsidies, and tradable quotas and permits), and other complementary policies (promoting research and education, family planning, modifications of trade policy, technology transfers). Most of these can have a domestic and/or an international dimension. Existing global climate change research that has analysed the range of policy instruments includes Mintzer (1988); US Congress, Office of Technology Assessment (1991); IPCC (1992);

National Academy of Sciences (1992); Clinton and Gore (1993); and McCann and Moss (1993).

Among the most recent studies on the climate change policy instruments, Heller (1997) provides a detailed analysis of the political economy of climate change and pays attention in particular to the treatment of normality of sub-optimality for the economies of transiting nations, such as China. He concludes that the formulation of an international climate change policy, and the definition of additionality in emissions trading systems, will err if it continues to treat sub-optimality as no more than a passing phase.

2.2.2   Carbon Sequestration and Non Energy GHG Emission Modelling

The literature focus on the potential for controlling CO_2 from energy sources is due in part to of the ready availability and adaptability of models that were designed to analyse energy markets. As regards to *carbon sequestration* and *non energy GHG emissions* there were no ready made macroeconomic models that could be adapted to analyse these activities. As a result the debates in these fields have not been structured around the bottom-up/top-down approach. Nevertheless controversies in these areas are related to issues very similar to the ones in the energy field; namely, the reasons for a wedge between the direct cost of technical alternatives from an engineering viewpoint and the overall cost of their adoption and implementation if transaction costs and economic general equilibrium effects are included in the accounting.

There are many difficulties in developing and comparing estimates of carbon sequestration costs. The differences among studies that give rise to the wedge between direct and social costs arise at each stage of carbon sequestration. They are the identification of forestry and agricultural practices and available land within the study region; the analysis of the costs of implementing the practices on the available land; and the analysis of the expected accomplishments of the practices and valuation of those accomplishments. Again, many factors can affect the estimates of carbon sequestration studies: including, data used with respect to land area, land costs, treatment costs, discount rates, carbon capture rates and patterns, ecosystem components and treatment of forest products.

To date, the studies fall into four geographic categories. One group is comprised of studies of the global potential and its cost of carbon sequestration (Sedjo and Solomon (1989); Nordhaus (1991b)). The second group considers the carbon sequestration potential of major ecological regions of the world (Dixon et al. (1991, 1991, 1994)). The third group concentrates on the potential of North America to sequester carbon (Adams et al. (1993); Moulton and Richards (1990); New York State (1991); van Kooten et al. (1992); Parks and Hardie (1995); Richards et al. (1993)). The fourth group, a set of recent studies, examines the potential for carbon sequestration in individual developing countries (Masera et al. (1994); Ravindranath and Somashekhar (1994); Xu (1994)).

A growing literature is emerging on non energy GHG emissions (methane (CH_4), nitrous oxide (N_2O), perfluorocarbons (PFC) and hydrofluorocarbons (HFC)). For example, the USEPA studies (1993, 1994).

2.2.3   Integrated Assessment Modelling

"Integrated assessment" is the convenient framework for combining knowledge from a wide range of disciplines. In fact, the integration process helps the analyst co-ordinate assumptions from different disciplines and introduces feedback absent in conclusions available from individual disciplinary fields. This means that the Integrated Assessment Models (*IAMs*) vary greatly with regard to their scope. The models that attempt to grapple with the full range of issues raised by the climate issue are referred to as *full scale* IAMs. There are three general types of integrated assessment models: i) models that seek to balance the *costs and benefits* of climate policies; ii) models that examine the implications of *targets* for climate change impacts; iii) models of sequential climate decision making under *uncertainty*.

The cost-benefit integrated assessment models focus on equilibrating the marginal costs of controlling GHGs and of adapting to any climate change. Any constraint on human activities is explicitly represented and costed out. At present, models of this type include very aggregate representation of climate damages (generally representing economic losses as a function of mean aggregate surface temperatures), sometimes disaggregated into market and non market damage components. Impacts on particular regions and sectors, which of interest of policy makers, are not explicitly represented in the current set of cost-benefit models. Early models of this type were also too complicated and it was difficult to incorporate explicit representation of uncertainty.

The targets-based IAMs provide details on the physical impacts of climate change on countries and regions in various market and non market sectors. Economic values have not generally yet been put on these impacts, reflecting both the paucity of valuation studies in some sectors and the modellers' perception that policy makers feel more comfortable trading off natural than economic impacts. Like the early cost-benefit models, these models have been too large to allow for sufficient sensitivity analysis and for explicit representation of uncertainty.

Reflecting the high level of uncertainty on the future evolution of socio-economic and natural systems, some analysts have begun to consider the climate change within explicit uncertainty frameworks. These models have generally either been the results of a relatively complete uncertainty representation of all key parameters within simplified models of the types discussed above, or the result of adding a limited number of alternative states to the full models discussed above.

The key components of full scale IAMs are four: human activities, atmospheric composition, climate and sea level, and ecosystems. Prior to 1992, only two integrated assessments of climate change models had appeared (Nordhaus, 1991a; Rotmans, 1990).

Some of the more common outputs of the cost-benefit type models are projections of the cost of controlling GHG emissions, of the damages resulting from climate change, of the *control rate* (stated in terms of the percentage reduction in GHG emissions in each year relative to level of emissions projected to occur in the absence of policy initiatives), and of the carbon tax required in each year to limit GHG emissions to the levels specified in the scenario under consideration. The cost-benefit decision for greenhouse policy involves a trade-off between substantial abatement costs early in the horizon, in return for avoidance of potentially large damages later. The discounting of future costs and benefits relative to current ones is a very critical value in such a trade-off (Nordhaus, 1994; Cline, 1992; Kolstad, 1993).

Targets type models tend to report land use by activity and/or physical impacts like ecosystems at risk, coastal land area lost, fresh water requirements, and mortality rates. Among these kinds of studies, there is the IPCC scenarios (ASF, from the USEPA, and IMAGE 1.0 models) of 1989, with analyses of delayed response, risk assessment, and cost-effective strategies for stabilising atmospheric CO_2 concentrations (Rotmans, 1990; Richels and Edmonds, 1993; Wigley et al., 1995).

The outcomes of the uncertainty oriented models are sensitivity analyses over key model inputs and parameters, and uncertainty analyses that focus on particularly relevant policy issues (Edmonds et al., 1994; Manne and Richels, 1993; Hope et al., 1993; Dowlatabadi et al., 1993; Nordhaus, 1994).

The costs of proposals to limit CO_2 emissions have been analysed by Richels et al. (1996) employing four widely-used energy-economy models (CETA, EPPA, MERGE, MiniCAM). They found that costs could be substantially reduced through international co-operation and the optimal timing of emission reductions. A summary of poll results coming from seven of the models participating in the Energy Modelling Forum Study 14, "Integrated Assessment of Climate Change" (CETA, DIAM, DICE, HCRA, MERGE, SLICE, YOHE) on the hedging strategies for global carbon dioxide abatement have been provided by Manne (1996a).

New results on the regional distribution of costs to stabilise atmospheric CO_2 under a hypothetical technology protocol have been provided by Edmonds and Wise (1997) using *MiniCAM* up to 2100. Manne and Richels (1997) used the *MERGE* model, to examine cost-effective strategies for limiting CO_2 concentrations and to explore the implications for near-term mitigation decisions and for long-term participation of developing countries.

Starting from the paper of Wigley, Richels and Edmonds (1996), Tol (1997) reconsiders the optimal timing of emissions abatement in an intergenerational equity context and demonstrates that removing the assumption of a long-lived decision maker with substantial foresight leads to higher near-term abatement.

The implications for developing (Non Annex I) countries of different emissions trajectories and protocols involving different concentration levels and time frames are examined by Shukla (1997). He considers two well known sets of trajectories, i.e. those proposed by IPCC (1995) and by Wigley, Richels and Edmonds (1996), and concludes that in the long-run the mitigation efforts in the developing countries will be crucial for the stabilisation of concentration.

The two region CETA-M model is used to explore some issues related to concentration targets by Peck and Teisberg (1997b). Firstly, they identify the cost and benefit assumptions that would make particular concentration ceilings optimal; secondly, they explore the acceptability to both regions of various burden-sharing agreements; lastly, they inquire whether the control effort and burden sharing rules may reasonably be negotiated independently.

The study of Montgomery et al. (1997) has two purposes. The first purpose is to describe a new model, called CETM, that combines a macroeconomic general equilibrium model of international trade with an engineering model of the energy sector; the second is to compare the results of CETM with those of MERGE on the effects of trade in non energy goods on the distribution of impacts between Annex I and Non Annex I countries when different emission limit scenarios are adopted.

Harrison and Rutherford (1997) deal with the coalition formation problem, mainly consisting in perceived inequity of the coalition and in incentives to free ride, and suggest an alternative solution. Their approach makes the global abatement decision one about efficiency, rather than perceived equity, and relies on the profit motive to bring about the efficient level of abatement. However, they demonstrate that equity concerns can still play a role in designing an attractive distribution of the burdens of abating global warming. The model that supports their results is based on the 25-region IIAM model developed by Bernstein, Montgomery, and Rutherford (1997).

Although integrated assessment of climate change is a rapidly evolving field, the following preliminary conclusions seem to emerge:

- integrated assessments are no stronger than the underlying natural and economic science that supports them;

- the IAMs show increased diversity in the distribution of regional costs and benefits;

- there are important gaps in research and inconsistencies between the information produced by the various disciplines whose reconciliation would bring marked improvements;

- while it is difficult to choose one policy in preference to others based on current knowledge about the climate system and human interactions with it, it has been demonstrated that the policy objective, the discount rate (see the recent theoretic contribution of Manne, 1996b) and the timing of compliance can be critical for short-term policy formulation and cost of action;

- given the considerable uncertainties associated with evolution of the climate system and its interactions with human activities, policies that enhance the flexibility of nations and individuals to respond, tend to have high value. in fact, R&D in technologies and institutions to facilitate the process of adaptation to global change generally have a high pay off;

- most current models do not match the social and economic organisation of the developing economies well. this can lead to biases in global assessments when impacts in the developing countries are valued as if their economies operate like those of the developed countries;

- climate change is but one dimension of global change; integrated assessments suggest that ecosystem impacts from projected climate change, agriculture management and urbanisation are of the same magnitude.

2.2.4 Costs as a Function of Baseline and Policy Strategies

The most sensitive issue is the way assumptions about the existence size of *no regret* strategies are conveyed in specific modelling frameworks and baseline scenarios. If the baseline scenario assumes the economy to be located on the theoretical production frontier, there is a direct and unavoidable trade-off between economic activity and the level of emissions. Conversely, in a baseline scenario, which describes an economy below the production frontier, *no regret* strategies are possible.

The existence of market and institutional failures that give rise to *no regret* potential is a necessary but not sufficient condition for the development of strategies to realise this potential. In practice, countries will consider climate policies in a multi-objective decision-making framework, whereby GHG mitigation policies are likely to be a by-product or a joint product of policies developed for other reasons.

Different underlying views of the efficiency of the economy correspond to the hypothesis discussed above. Since many existing models can adopt either view of the economy, such underlying assumptions are often the main reason for the differences in quantitative results among different analyses.

2.2.5   Socio-economic Assumptions Underlying Scenarios

Most models explore options, alternatives, and possible future economic conditions, based upon particular assumptions about the nature of technological change and long term development paths.

GHG emissions over the long run depend not only on the rate of economic growth but also on the structure and physical content of this growth. Countries at a similar development stage may have very different energy consumption per capita ratio, or very different transportation requirements. Comparative studies aiming at explaining these differences (Martin, 1992; Darmstadter et al., 1977) suggest the importance of five considerations: i) technological patterns in sectors such as energy, transport, heavy industry, construction, agriculture and forestry; ii) consumption patterns; iii) geographical distribution of activities; iv) structural changes in the production system and in particular the role of high or low energy intensive industries and services; v) trade patterns (it is generally argued that removing tariff and non tariff barriers enhances overall economic efficiencies).

These factors are not ignored by current economic models: they are in some way captured by changes in economic parameters over the short term (1 to 3 years) and the medium term (4 to 10 years). For the longer term the economic parameters cannot easily be viewed as the sole command variables used to predict the future of production and consumption systems. The outcome in terms of GHGs will also depend upon dynamic linkages between technology, consumption patterns, transportation and urban infrastructure, urban planning or rural-urban distribution of population.

The mitigation costs attached to each possible baseline scenario depend not only on the absolute level of the required mitigation and on the array of available technologies but also on the timing of this mitigation (Grubb et al., 1993b; Hourcade and Chapuis, 1993; Manne and Richels, 1992; Richels et al., 1996). In this connection, three issues become very important: i) the flexibility (inertia) of consumption patterns underlying the activity of GHGs emitting sectors; ii) the behavioural characteristics which determine technical change and the evolution of life style; iii) the interactive effects due to the feedback between the use of certain options and the rest of the economy.

Moreover, four factors are likely to give rise to different development paths underlying different baseline emission levels.

1. Material and energy content of development in industrialised countries (over the past several decades, the raw material intensity of industrialised countries has dropped significantly; IAEE, 1993). The major socio-economic factors that will affect the future raw materials intensity are: structural shifts in the economy towards services; increases in the information-intensity of industrial processes, goods and services (Chen, 1994); the effects of telecommunications on travel and transport energy use (Selvanathan and Selvanathan, 1994);

saturation in the consumption of some goods and services, and the emergence of less energy and material intensive goods and services.

2. The links among energy, transport and urban planning. Transportation energy use accounts for a significant proportion of GHG emissions and its growth rate is typically higher than for other categories of energy demand. In the transport sector, the size and type of GHG emissions are a function of the demand for transportation services, the mode chosen, the efficiency of the vehicles and the types of fuel used.

3. Land use and human settlements. This issue concerns all the regions of the world but its quantitative impact is more impressive in developing countries. Land use and human settlement pattern changes derived from agricultural and forestry activities as well as rural-rural and rural-urban migrations are among the main sources of GHGs in these countries. In this field it is even more important than elsewhere to look at the GHG mitigation component of more general development strategies: mitigation costs are only one part of a larger set of factors in such cases.

4. Development patterns in developing countries. As a major part of the needed infrastructure to meet development needs is still to be built, the spectrum of future options is considerably wider than in industrialised countries. The traditional approach of using *business as usual* assumptions as the baseline is then particularly problematic.

2.3 Gaps from Previous Studies and Directions of Future Research

The extensive literature already devoted to GHG abatement has helped identify the main aspects of the climate debate issues. From the point of view of the EU and from a methodological viewpoint, four main issues still remain open.

- The cost of GHG abatement for the medium short term at the level of individual EU Member State. Most of the studies with economic models conducted so far treated Europe as one region, or did not include all member states and did not include sufficient sectoral disaggregation.

- The models do not start from the observed social accounting matrices of EU countries, that would make possible the identification of existing levels of taxation and the mechanisms of income re-distribution.

- While there are many studies on the cost of emission abatement, the benefits from a better environment remain extremely difficult to quantify. A recent project financed by the European Commission, called EXTERNE has attempted to evaluate environmental damages.

- The issue of how technological development can be shaped by public policy as well as the hedging strategies for long term energy technologies are issues virtually not covered by empirical work. They have only been partly covered in models studying the very long term, but using reduced form and very aggregate models.

The simulations in the present volume with the GEM-E3 model attempt to cover the first two issues and present some first findings on the third. The model starts from a detailed description of the initial situation of the EU economies[14]. It uses the EXTERNE data to quantify environmental quality (in one scenario external costs are internalised in the economy). It also gives some preliminary results on the incorporation of endogenous technology dynamics as part of the decision process of economic agents.

---

[14] For details on the GEM-E3 model and its database the user should refer to the GEM-E3 model reference manual.

3 Definition of Policy Instruments and Criteria for their Assessment

3.1 Introduction

Curbing GHG emissions could invoke a considerable burden on the economy. Consequently, any environmental policy plan must rely on a careful design of the instruments through which the environmental goal is to be reached. Given the variety of instruments that can be considered, it is necessary to categorise them in a systematic way. A unified set of criteria for the evaluation of the effectiveness of these instruments must also be compiled.

The implementation of environmental policies can be achieved through a variety of policy instruments. The choice of instruments will depend on a set of criteria and on the type of environmental problem to be tackled. In some cases, a mix of instruments may be required. In other cases, certain instruments may be more suitable for tackling specific environmental problems.

Whereas during the last two decades the regulatory approach dominated national as well as EU-wide environmental policies, there seems to be a rising trend in the use of economic instruments, including environmentally-motivated taxes on products or emissions and deposit-refund systems (OECD 1997). This increasing acceptance can be attributed largely to the growing awareness of serious environmental and economic deficiencies of the traditional command-and-control regulation system, but also to the need for reduced regulatory complexity.

In accordance with this general trend, the European Commission has started moving towards a new environmental policy concept that is more in line with the market system than the traditional command-and-control legislation. Thus, in the Community Fifth Environmental Action Programme and its 1996 review, the Commission has followed a policy approach and strategy which differs decisively from previous programmes. As indicated by the programme's title "Towards

Sustainability", longer term objectives are formulated and a more global approach has been developed. The Commission stresses, among other things, the necessity to design improved approaches to climate change and mentions explicitly the need to broaden the range of instruments by market-based, horizontal and financial instruments in order to enable the effective implementation of the programme until 2000.

In addition, the new strategy outlined in the programme is oriented towards a greater co-operation, which provides that all economic agents and social partners have to be involved in the policy making process. This is in line with the Commission's Communication (COM (96) 561 final) on environmental agreements, where general guidelines for the effective and acceptable use of environmental agreements have been set out for discussion.

In the following sections selected policy instruments will be described and evaluated with respect to a set of criteria following the concerns of policy makers. The focus of this review lies on traditional regulatory instruments, environmental taxes, tradable permits, and voluntary agreements. The emphasis is on air pollutants, especially CO_2 but also SO_2, NOx, VOC, particulates and ozone as energy use is one of their major sources.

3.2 Criteria for the Assessment of Environmental Policy Instruments

Designing an optimal policy that maximises net benefits requires the resolution of a number of problems related to both fundamental assumptions and practical application aspects (Klaassen 1996a, pp. 18ff.). These include:

- information problems (concerning the environmental damage)
- distribution problems (concerning the balance between equity and optimality), and
- ecological problems (the pollution level achieved may exceed the assimilative capacity).

The effectiveness of environmental policy instruments can be evaluated by *cost-benefit or welfare analysis* (maximising the net benefit of a policy) or by using the less strict criteria of cost-efficiency. This last approach selects the instrument to meet a politically given ecological standard at lowest cost, taking into account distributional effects and uncertainty (Klaassen 1996a, p. 19).

Figure 3-1 lists some relevant criteria for the measurement of the effectiveness of environmental policy instruments, which are widely accepted in the literature (Klaassen 1996a, Bohm and Russell 1985, pp. 399ff, Siebert 1992, pp. 128ff.).

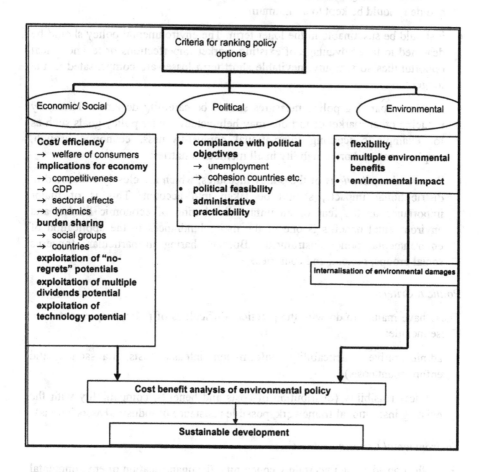

Figure 3-1: Criteria for ranking policy options.

Three groups of criteria are distinguished with respect to economic, social and environmental aspects.

Economic Criteria

Being an additional constraint on the economy, any environmental policy is likely to induce adverse effects on the economy. Curbing GHG emissions may affect all economic agents. Given its global perspective, it should be designed in such a way so that it complies with economic and social goals that are considered important. Any policy mix should therefore meet the following criteria:

- it should be designed in a cost-effective way i.e. the effort per ton of carbon avoided should be kept to a minimum;

- it should be sustainable in the loner term. The environmental policy should be designed to take advantage of existing market imperfections or technological opportunities so that any inevitable short-term losses are compensated in the longer term;

- any accompanying policy measures should be carefully designed. These, by reducing other market distortions may help achieve other policy goals such as for example concerning international competitiveness, economic growth, employment, and price stability in all member countries;

- the *social implications* of the policy measures, which are closely connected to distributional impacts, should be taken into account. This is of major importance as the fear of an unfair distribution of economic burden and environmental benefits is one of the major hindrances in the application of environmental policy instruments. Burden sharing in particular, concerns social groups, sectors, and countries.

Political Criteria

These have mainly to do with the practical difficulties of policy implementation. These include:

- administrative practicability (information intensity/costs, transaction and enforcement costs)

- political feasibility (distribution of costs and benefits, compatibility with the existing institutional framework, possible resistance of industrial associations)

Environmental Criteria

Given the considerable uncertainty concerning the quantification of environmental damages and the long horizons involved, the environmental policy should aim at:

- flexibility with respect to changing circumstances (ability of the implementation system to adjust so as to keep the environmental goal even when exogenous shocks - in costs, technologies, economic activities, and the price level (inflation) - occur (Bohm and Russell 1985);

- ability to generate multiple environmental benefits related to other pollutants.

Environmental policy must also take into account elements such as the characteristics of the environmental problem, its monitoring ability, the uncertainty component both in benefits and abatement costs, the market structure/market power in the sectors to be controlled. The spatial dimension is important in the case of GHGs, which is a global problem. The accumulation of greenhouse gases in the atmosphere is not related to the geographic source of emissions: greenhouse gas is a "uniformly mixed pollutant". For other problems, like acid rain, the source of emission, and the place of deposition matter, i.e. the pollutants are non uniformly mixed. *Greenhouse gases are cumulative pollutants, like acid rain, their stock dimension must be taken into account in the policy design.* GHG like, many environmental issues have international implications and will require co-operation between countries, through bilateral or multilateral agreements. This can strongly influence the choice of instruments (Segerson 1995).

3.3 Selected Environmental Policy Instruments

Three groups of instruments can be distinguished:

- regulatory or command-and-control instruments;
- market-based instruments;
- other, including voluntary agreements, informational and organisational instruments.

Within the market-based instrument category, one can distinguish charges and taxes, subsidies and tradable permits. These instruments can be used unilaterally by individual nations, by a group of countries or in compliance with a multilateral agreement. Table 3-1 gives an overview of the set of alternatives environmental policy instruments the policy maker can use.

Table 3-1: Classification of policy instruments.

Domestic and International Policy Instruments		
Regulatory instruments	Market-based instruments	Other instruments/policies
• performance standards • technology standards • product standards • product bans	• taxes and charges domestic tax international tax internationally harmonized tax • tradable permits domestic tradable permits • international tradable quota system • joint implementation • subsidies	• informational and organisational instruments (e.g. eco-audits, eco-management, R&D programmes) • other complementary policies (e.g. public education, family planning)
• voluntary agreements (may include regulatory as well as market-based or other instruments)		
	• environmental liability law	

Policy instruments of particular reference for combating climate change at the international level are tradable emission quotas/permits[15], joint implementation, internationally harmonized domestic taxes, international taxes, non tradable quotas/permits, international standards or voluntary agreements. At the national level, countries can choose between domestic taxes and tradable permits, technology and performance standards, product bans, subsidies, voluntary agreements or other instruments, like public education.

3.3.1 Regulatory Instruments

Direct regulations can be described as "institutional measures aimed at directly influencing the environmental performance of polluters by regulating processes or products used, by abandoning or limiting the discharge of certain pollutants, and/or by restricting activities to certain times, areas etc. Their main feature is,

---

[15] Frequently, for instance in the IPCC report of Working Group III, the term 'tradable quotas' is used for international instruments, whereas the term 'tradable permits' is reserved for domestic instruments (Goldemberg et al. 1996, p. 37).

that a specific level of pollution/abatement is prescribed" (OECD 1994, p. 15). The regulatory concept therefore aims at attaining a given environmental target by regulating individual behaviour (Siebert 1992, p. 131).

The conventional range of instruments for environmental policies includes technology and performance-based standards or product standards or bans (e.g. fuel standards, ban of aerosol sprays). Technology-based standards typically require the use of specified equipment, processes, or procedures (e.g. best available technology). Performance-based standards allow more flexibility since they prescribe an admissible level of emissions or polluting activities but no means how to realise it (e.g. levels of energy efficiency for electrical appliances) (Fisher et al. 1996, p. 412). The precise way standards are defined is of great relevance for the cost and the performance of this instrument (Helfand 1991). Optimising the implementation of a standard can be more important than switching to another category of instrument.

Wide experience with the application of emission standards, combined with technology standards, has been gained in air-quality policy in Europe, as well as in the United States and Japan (Siebert 1992, p. 131). These standards constitute the majority of environmental policy instruments adopted in Europe. For instance, the reduction of SO_2 emissions during the last ten years has been brought about by the installation of end-of-pipe technologies (flue-gas desulphurization) in power stations under the European Directive for Large Combustion Plants of 1988 and by standards on the maximum sulphur content of fuels.

Regarding the CO_2 target, uniform standards have been used for mainly in areas such as energy efficiency of buildings, fuel use of motor vehicles, energy efficiency of household appliances, and the emission content of fuels (Fisher et al. 1996, p. 403).

3.3.2 Environmental Taxes and Charges

In accordance with the Commission's definition[16], "taxes and charges" are understood to be a generic term for "all compulsory, unrequited payments, whether the revenue accrues directly to the Government budget or is destined for particular purposes" (COM (97) 9, p. 3). The term environmental taxes is used to subsume all taxes and charges which tackle environmental issues and aim at reducing environmental degradation.

The application of environmentally motivated taxes is based on the assumption that environmental problems are partially due to a failure of the market system. Taxes and charges which are imposed on activities that have a negative impact on

---

[16] Set out in the Communication on Environmental Taxes and Charges in the Single Market of January 1997.

the environment. They take over the signal and control function of (missing) market prices and incorporate the environmental costs in the price of products and services. A CO_2 tax can be interpreted as the price for the use of the global atmosphere.

Basically, three functions of environmental taxes, which frequently appear together, can be identified (OECD 1997, p. 16):

- incentive function,

- earmarked (or revenue-raising) function,

- fiscal function (eco-taxes).

Taxes with an *incentive function* aim at correcting the price signals to producers and consumers in order to reflect the cost of environmental utilisation. Two basic theoretical approaches for determining the level of the taxes can be recognised, one developed by Pigou and the other by Baumol and Oates. Pigou (1920) starts from the assumption of a market failure due to external effects. An optimal internalisation of externalities can be realised if the price system is complemented by "ecological right prices" reflecting the social costs of environmental utilisation. In neo-classical welfare theory, the optimal tax rate has to be set equal to the marginal external damage. Hence, the Pigovian concept is an optimisation approach that determines the environmental quality endogenously. Unfortunately, in reality these are information gaps on the damages caused by economic activities and their monetary valuation.

Steering taxes, following the "charges and standards" approach proposed by Baumol and Oates in the 1970s, are mainly directed towards achieving specified emission standards at minimum abatement costs to society. In this case "one gives up any attempt to reach the true social optimum"; the target variable is the outcome of a political process (Baumol and Oates 1988, p. 155). Taxes are used to reach an environmental target at minimum cost.

Taxes with *earmarked function* are imposed primarily in order to finance pollution control measures, such as the clean up of disused landfill sites. In this case the determination of the tax rate and its base is oriented more towards the revenue target rather than towards an emission reduction target.

The OECD (1997, p. 16) has defined a new category of environmental taxes with a *fiscal function*, the eco-taxes. Eco-taxes differ from taxes with earmarked function with respect to the allocation of the revenues. They differ from incentive charges as well, because the primary function of the tax is fiscal (i.e. shifting of tax burden by replacing non environmental taxes) and not environmental. Because they have an environmental tax base, they keep an incentive function, but the revenues are used in order to finance cuts in existing distortionary, non environmental taxes.

For uniformly mixed pollutants, which involve damages at world level, like CO_2, a broader spatial coverage of a tax has to be considered. As Table 3-1 points out, three types of tax can be distinguished (Muller 1996, pp. 470ff., Hoel 1992, pp. 101ff., Fisher et al. 1996, pp. 417ff.):

- domestic taxes,
- an international tax, as well as
- internationally harmonized domestic taxes.

A *domestic tax* can be introduced by the domestic government in order to reach a national emission reduction target, regardless of whether the target is determined by national policy or an international agreement. The tax rate is specified and the tax revenue is collected and used by the domestic government. A domestic carbon and/or energy tax leads to higher energy prices and to an overall price increase. This may - in the short term - have negative impacts on the international competitiveness of domestic firms. Highly energy-intensive branches may lose market share in the world market as well as, in the absence of border tax adjustments, in the domestic market. In order to prevent serious damage to domestic producers, provisions for industry exemptions from the CO_2 tax, or lower tax rates for highly energy-intensive sectors respectively, have been proposed. Tax exemptions are granted in Denmark, Norway and Sweden (Muller 1996, p. 473)[17] The exemption of specific sectors from tax gives ecologically negative signals and may shift demand towards the products of these sectors. Ideally, competitiveness should be avoided through the use of a global tax or a multilateral agreement harmonising national taxes.

An *international tax* system may be based on an international greenhouse gas agreement, through which, the participating countries agree on a uniform tax rate for CO_2 emissions and on some revenue distribution rules. The determination of a redistribution rule under an international tax system is equivalent to the determination of the initial allocation of quotas under an international permit system. Under an international tax the countries themselves are free to choose their own domestic policy instruments as well as their own emission reduction targets. As Hoel (1992) argued, under ideal conditions, i.e. no market failures and distortions in existing markets, it would be optimal for the government of each country to impose a domestic tax with a tax rate equal to the international tax rate. In practice, the implementation of an international tax faces a number of problems, concerning the establishment of an institutional framework (including an international authority) and an effective enforcement regime. Hoel (1992, p. 109)

---

[17] The Commission points out in its Communication on Environmental Taxes in the Single Market (COM(97) 9) that exemptions from emission or product taxes may constitute state aid in the meaning of Art. 92(1) of the EC Treaty. This requires that the tax relief or compensation is temporary and should not provide the sector exempted with a net benefit.

has shown that an agreement of harmonized domestic taxes seems to be potentially more realistic and promising in the near future.

Within *multilateral agreements to harmonise domestic taxes* the national tax rates are determined as part of an international agreement. In contrast to international taxes the tax revenue is collected and redistributed by the national governments. Harmonised domestic taxes are similar to an international tax if the domestic responses to the international tax are uniform domestic taxes. The reimbursement rule is designed in such a way, that in equilibrium each country pays zero net taxes. Problems with an international agreement of harmonising domestic taxes include the free rider problem and the difficulty of taking into account existing taxes on fossil fuels or other policy measures affecting CO_2 emissions. Regarding the latter aspect Hoel (1992, pp. 106ff.) draws the conclusion that the only practical solution is a uniform minimum CO_2 tax for all countries. It then should be up to the individual countries to decide, whether the tax will be added to or whether it should replace existing taxes.

Examples for multilateral agreements to harmonise national taxes are the Commission's proposal for a EU-wide carbon/tax, presented to the Council of Ministers as a draft legal directive in 1992 (COM (92) 226 final), as well as the recent Commission's proposal on a common EU-wide excise tax duty system on energy (Europe Environment 1997a, pp. 4ff.). Both proposals address the issue of European competitiveness vis-à-vis third countries, e.g. the USA or Japan. The first by allowing graduated tax reductions for energy-intensive branches down to full exemptions of firms that are investing in CO_2 reduction measures and the second by suggesting a refund system for very energy-intensive firms. Whereas the EU proposal on the energy/CO_2 tax has followed the principle of 'conditionality', i.e. an implementation in the EU depends on similar measures taken at the international (OECD) level, the recent Commission's proposal for an excise tax system does not include a conditionality clause and it was criticised by part of the industry. The affected industries - represented by the Union of Industrial and Employers' Confederations (UNICE) fear that the proposal would threaten the competitiveness of European firms without being cost-effective in terms of its environmental objectives (Europe Environment 1997b, p. 1).

Whereas in theory a multilateral tax or a harmonized national tax system are superior to national taxes, experience shows that in practice such multilateral approaches have not been very successful (Muller 1996). In order to support national initiatives, the legal framework for the introduction of environmental taxes, set by the EU and WTO in particular in the field of border tax adjustments, has to be clarified and barriers of unilateral actions have to be reduced. First steps were undertaken by the European Commission in setting out guidelines for Member States in designing, implementing and evaluating environmental taxes (COM (97) 9).

There are examples of successful international agreements on global or transboundary pollutants, like the Toronto agreement on the ban of CFCs or the Helsinki agreement on SO_2 emissions. These agreements allow countries full flexibility on how reach the agreed targets.

Information regarding the use of economic instruments in OECD countries (OECD 1994) is presented in the following table.

Table 3-2: Charges and taxes in practice.

Instrument	Objective	Comments
Charges on emissions	Control of air and noise pollution	The tax revenue is applied to the general budget, air quality control, compensation of health damage, noise abatement, or is rebated to energy producers
Charges on products a. Car sales tax differentials b. Differentiation of annual vehicle tax c. Tax differentiation between leaded and unleaded petrol d. Charges on fossil fuels (carbon tax and charges on sulphur content of fuels) e. Other product charges	a. Control of air pollution by means of new technologies and fuel efficiency b. Control of air pollution by means of new technologies c. Control of lead and other air pollutants' emissions d. Control of CO_2 and SO_2 emissions e. Control of air pollution in diesel and gasoline cars and on domestic aviation, and reduction of electricity consumption in the residential sector	a. Taxes are based on the performance of cars regarding emissions, weight, fuel efficiency, use of alternative fuels and new technologies b. In most cases tax revenue is applied for the general budget. It is also applied to road construction and maintenance
Permits	Control of air and water pollution	Tradeable permit systems are used in the USA, in Canada and Australia. They were also implemented in Germany, though not very successfully

3.3.3   Tradable Permits

Tradable permits, similar to emission taxes, make use of market mechanisms. A trading system for pollution rights can be designed in several ways depending on the ecological characteristics of the pollutant being controlled (see below). The simplest case is a system of locally undifferentiated emission permits. For a regional emission standard of pollutants, which are uniformly mixed in the atmosphere, such as CO_2, this type of permit system provides a maximum of economic efficiency and market functioning. It is fully equivalent to a theoretically first-best ambient permit system from an economic as well as from an ecological point of view (see below). For non uniformly mixed pollutants (for which the spatial distribution as well as the total amount of emission matters), the design of a permit market is far more complicated as a separate market for each receptor has to be defined.

A permit system consists of three elements (Koutstaal and Nentjes 1995, p. 220): an ecological element; a distributive element; and a an economic element.

The *ecological element* concerns a politically determined amount of total emissions for a region. This has to be determined by an environmental control authority and divided into a number of emission rights (permits). An emission permit is defined in terms of an allowable emission rate per year; it can be time-limited or valid for an unlimited time period.

The *distributive element* considers the initial allocation of permits. For example, permits could be distributed free of charge, according to emission levels in the past (grandfathering) or through an auction. The possibility of grandfathering is an attractive feature for existing emitters, as it reduces additional financial burdens, in particular for energy-intensive industries. As firms have to bear abatement costs only, resistance on the part of affected industries may be less strong than with a tax regime, where firms are burdened with abatement costs plus tax payments for residual emissions. Depending on the reference chosen to determine the initial allocation, the grandfathering approach may lead to unjustified preferential treatment of old firms, discrimination of new firms and an ecologically undesired slowing down of technical change, especially if the permits are time-unlimited. Nevertheless, the grandfathering approach turns out to be the main allocation mechanism due to political acceptability reasons. For instance, in Title IV of the 1990 amendments to the U.S. Clean Air Act, the so-called Acid Rain Program, grandfathering was chosen. Within the Acid Rain Program, the baseline to determine the allowance allocation was given by the average annual fuel consumption from 1985 to 1987. Using targeted SO_2 emission standards as reference allocation avoids the problem of rewarding laggards by granting them more valuable pollution rights than early emission reducers. Grandfathering has been supplemented with an auctioning to allow for a more equitable treatment of new entrants (Koschel and Stronzik 1996).

The *economic element* concerns the efficiency effect induced by trade. The theoretical functioning of tradable permits can be explained by the following mechanism. Within an emission permit system the firms can choose between:

- purchase of tradable permits (or renunciation of selling permits respectively) and emission of the corresponding amount of pollutants or

- investing in new emission control technologies and renunciation of purchase of permits (or sale of unused permits respectively).

Every firm individually decides whether it is more efficient to buy a permit or invest in additional improved control technologies. Whereas a policy of technical standards reduces the firm's private autonomy and its flexibility of technology choice considerably, a policy of emission permits allows maximum freedom in the individual adjustment strategy.

3.3.4   Subsidies

It is questionable whether subsidies should be classified as market based instruments. Subsidies might serve to support the polluter-pays principle by aiming to overcome the financial bottlenecks, which impede expeditious implementation of environmental protection initiatives. Theoretically the same effect as with taxes can be achieved if subsidies are linked to emission reduction (Cansier 1993, pp. 140ff.). This theoretical model, however, is open to criticism due to the system's high information requirement and the long-term economic inefficiency caused by transmission of incorrect price signals. Through subsidisation, the price of the pollution-intensive good is too low and therefore the price structure as an allocation guideline is distorted (Siebert 1992, pp.130f.).

A linked subsidy/tax system has also been discussed e.g. by the IPCC. Here, all polluters face a tax for units of emissions above an allowed (firm-specific) baseline, whereas they receive a subsidy for reductions in emissions below that baseline. One disadvantage of such 'carrot and stick' systems is the administrative cost associated with the setting of a firm-specific baseline (Fisher et al. 1996, p. 403). The main task of environmental subsidies should rather be to supplement and support the range of existing regulatory instruments. In the transitional phase of environmental policy, towards wider application of polluter-pays instruments, the use of state-financed promotional programmes could give incentives for sophisticated avoidance activities and could help to eliminate information deficits (Hansmeyer and Schneider 1990, p. 53). Fuel subsidies, which have been introduced in some countries for distributional reasons, have negative effects on economic efficiency. A number of studies have pointed out that global emissions would decrease by 4-18%, if fuel subsidies were abolished (IPCC 1996, p. 15).

3.3.5   Voluntary Agreements

Voluntary agreements can take the form of unilateral declarations on the part of industry or bilateral commitments between the public authority and industry (usually a business association or an industrial federation). The industry promises to reach an (often sector-based) set of environmental objectives according to a schedule included in the agreement. There is no formal commitment on the part of the authority but often promises are given to exempt the sector involved from alternative regulations concerning the same environmental problem (Glachant 1994, p. 38). Thus, voluntary agreements often are designed to avert the enactment of stricter laws or statutory ordinances, which are believed to be more costly to the industry (Kloepfer 1991, p. 740). The industry gains from voluntary agreements if compliance costs of the agreement are lower than compliance costs of the alternative instrument threatened.

The effectiveness of voluntary agreements depends decisively on their legal status. They can be designed in a legally binding form, with formal sanctions in case of non compliance, or in a non binding form.

Although voluntary agreements were used already in the 1970s, they have become more common in recent years. Since the beginning of the 1990s, an upward trend towards more self-regulating instruments and less governmental intervention has been observed. More than 300 agreements have been concluded at national level in the EU Member States (Dröll 1997). Though more common in the fields of waste-management and phasing-out of chlorofluorcarbon, voluntary actions for energy-related CO_2 abatement have appeared over the last five years (IEA/OECD 1995). As Table 3-3 demonstrates, voluntary agreements in the area of CO_2 typically aim at increasing energy efficiency. In contrast to the increasing interest in voluntary agreements for environmental protection, the theoretical analysis of this instrument is rather weak (Carraro and Siniscalco 1996, p. 80).

In Germany, for instance, the industry has issued about 80 environmental agreements since the late 1970s. Most of these are unilateral declarations in a non binding form and without legal status. Recent examples of voluntary agreements in Germany include the take-back agreement for cars, the agreement made by the automobile industry on the development of energy efficient cars and the voluntary agreement made by a number of industries on a CO_2 reduction by the year 2005 (undertaken in view of the 1995 Berlin Climate Conference) (see Rennings et al. 1997 for further details). In reaction to the ambitious CO_2-reduction policy of the Federal Government (25 % by the year 2005 compared with the 1990 level) the German business community presented a non binding voluntary agreement in March 1996, which is an updated version of two previous agreements of 1991 and 1995 (Rennings et al. 1997, Breier 1997). In return for the industry's commitment to reduce CO_2 emissions, the federal government of Germany promised to abstain from the introduction of other regulations, such as the 'Wärmeschutzverordnung' (Building Insulation Code). The agreement foresees an annual monitoring of the reached emission reductions. While compliance seems not to be the major problem, there are some doubts that the target stated in the agreement will be sufficient to reach the national 25 % reduction goal.

Table 3-3: Experiences with voluntary agreements/actions to reduce CO_2 emissions in the EU member countries.

Country	Number of voluntary agreements	Voluntary actions for energy-related CO_2 abatement (selection)
Belgium	14 (since the late 1980)	• energy sector: use of combined cycle power plants, promotion of cogeneration of heat and power • industrial sector: energy saving, support of energy audits and energy management • households: energy efficiency standards for electric appliances
Denmark	16 (since 1987)	• energy sector: use of renewable energies • industrial sector: agreements with NGOs concerning measures for increasing energy efficiency
Germany	80 (since the late 1970's)	• energy and industrial sector: 19 branches have committed to reduce their specific CO_2 emissions by 20% until 2005 (base year 1990); 12 industrial branches have committed to absolute reductions (80% of the energy consumption by the German industry is covered by this revised commitment)[18] • transport sector: 25% reduction of specific fuel consumption of new cars from1990 to 2005
Spain	6	• energy sector: promotion of fuel substitution, cogeneration power plants and renewable energies • industrial sector: energy saving

---

[18] Bundesverband der Deutschen Industrie (1996).

Table 3–3: Continued.

France	about 20	• energy sector: support of renewable energies • industrial sector: about 10 agreements concerning CO_2 emissions from industrial sectors (smelting, chemical, paper, welding, glass, plaster, cement) are being prepared; an agreement concerning CO_2 with the aluminium sector has recently been signed • households: promotion of energy efficient electrical appliances • transport sector: support of vehicles based on electricity or gas
Greece		• industrial sector: energy audits, investment in energy saving measures
Ireland		• industrial sector: audits and firm declarations on energy balances • households: labelling of refrigerators and washing-machines
Italy	11	
Luxembourg	5	
The Netherlands	more than 100	• energy sector: programmes on energy efficiency, NO_x reduction, use of cogeneration power plants and renewable energies • industrial sector: energy saving targets • households: insulation, energy-efficient heating and lighting, energy-efficient refrigerators freezers
Austria	25	• energy sector: support of renewable energies • industrial sectors: energy audits for industrial energy consumers
Portugal	10	• industrial sector: energy saving measures
Finland		• industrial sector: agreements with processing industry branches concerning energy saving • public sector: saving of electricity in public buildings, fuel saving

Table 3-3: Continued.

Sweden	13	• energy sector: support of cogeneration power plants based on biomass • industrial sector: individual commitments of firms to enhance energy efficiency • households: efficiency standards for electrical appliances
UK	8	• energy sector: targets concerning the extension of cogeneration capacity • industrial sectors: individual commitments of firms to enhance energy efficiency • households: labelling of refrigerators and freezers

Source: COM (96) 561 final, Annex, and IEA/OECD (1995), own compilation.

In the Fifth Environmental Action Programme of the Community a new approach of co-operation between industry and public authorities, including the application of voluntary agreements and other self-regulation, has been developed. According to a Communication on Environmental Agreements on the EU level the Commission presented in November 199, the general purpose of the Communication is to promote and facilitate the use of effective and acceptable Environmental Agreements" (COM(96) 561 final, p. 5). Due to the responsibilities of the European Community, the Communication mainly addresses agreements concluded at Community level and national agreements used to implement Community directives.

In the general guidelines, set out in the Communication, the Commission stresses:

• the necessity of a legally binding form

• prior consultation with interested parties

• quantified and staged objectives

• the monitoring and report of results, as well as

• the publication of the agreement and of the results obtained

In addition, one objective of the Communication is to clarify the suitability of voluntary agreements as an instrument for implementing environmental objectives at the European level set by Community directives.

3.3.6　Environmental Liability Legislation

Environmental liability legislation constitutes an internalisation strategy, which leads, under ideal conditions, to an optimal avoidance of damages by forcing the polluter to pay for any damage he causes (see e.g. Endres 1994, Cansier 1993). Liability legislation may supplement other environmental policy measures in areas where the state lacks information on future environmental damages or where the target is solely the avoidance of liability-relevant reparable damage (SRU 1994, Tz. 567).

Two basic liability principles can be distinguished. Under default liability, the incentives are limited to compliance with the statutory standard for due care and diligence. Under the regime of absolute liability, there may be, additional, incentives to avoid damage arising from emissions in normal operation. However, the absolute liability regime cannot be utilised for implementing politically specified emission reduction targets, since the degree of environmental quality will emerge endogenously from the market itself. In reality, the ecological efficacy of liability legislation will be severely restricted the case of CO_2 emissions because these are cumulative and long-distance damage categories, not permitting assignment to particular polluters involved. When it comes to preventive environmental protection, mandatory liability and insurance are completely unsuitable options.

3.4　Evaluation of Environmental Policy Instruments

The following evaluation of instruments will be based on the following set of criteria (see Figure 3-1):

- environmental effectiveness
- static cost efficiency
- dynamic efficiency
- non environmental benefits
- administrative practicability, and
- political feasibility

3.4.1　Environmental Effectiveness

In theory, *direct regulations* represent the most favoured option if emissions have to be reduced in the immediate vicinity of particular facilities and if for ecological reasons relocational reactions by the operators involved are seen as undesirable. In the case of very harmful and locally concentrated emissions no margin of error can be accepted, the inflexibility under a regulatory system constitutes an important advantage over more flexible economic instrument. If enforcement is effective, standards provide complete compliance with the goals established. However, the attractiveness of the regulatory approach diminishes if enforcement or implementation is weak. Nevertheless, the use of direct regulation is certainly recommended when effectiveness, i.e. reaching the environmental target, is the main decision criteria, as is the case of toxic elements.

In the case of risk prevention where no actual danger exists and a correlation between cause and effect has not yet been significantly proven - as in the case of global warming - there is no necessity to use regulatory instruments. However, if environmental policy aims at reducing the total quantity of emissions facility-specific regulatory instruments can be less effective than economic instruments which affect total emissions. Emission standards for individual firms, for instance, can not guarantee that a given total amount of emissions will not be exceeded. In practice, e.g. in Germany, emissions can be reduced through the imposition of technical standards like the 'best available technology'. But there are no economically or legally well-founded reasons why the regulatory approach should be the best option for precautionary action. In fact, the implementation of the precautionary protection provides a range of opportunities for the application of economic instruments, which are more appropriate to support the long-term development of new and environmentally friendly technologies.

In principle, solutions involving economic instruments provide a higher degree of individual freedom and flexibility, but they may delay the prompt elimination of an environmental risk. The comparative advantage of *environmental taxes* (and tradable permits as well) is primarily in the field of preventive environmental protection. Whereas tradable permits (given appropriate spatial differentiation) will guarantee precise compliance with quantified ecological targets, taxes are less direct instruments if a regional emission limit has to be met, as information gaps regarding abatement costs may lead to setting on inappropriate tax level.

Adjustments of the tax rate according to a trial-and-error procedure may lead to delay in meeting the environmental goal. There is also a danger the emission tax will overshoot and lead to unnecessarily high costs of attaining the target (Siebert 1992, p. 122). Therefore, emission taxes are more suitable for those types of pollution for which the avoidance of high costs is much more important than the attainment of an environmental target in the short term (Klaassen 1996a, p. 23). In practical terms, taxes are frequently applied together with direct regulations in order to combine the reliability of emission standards and the incentives of a tax system (Siebert 1992, p. 148). An example for Germany, is the so-called "residual pollution charge" in the statutory water code.

A major disadvantage of taxes is that, in contrast to tradable permits (see below), they have to be adapted to changes in the economic conditions (economic activity, inflation, cost-saving technical progress) in order to reach a predetermined environmental target. In summary, the information intensity for implementation of environmental taxes is comparatively high. It can be reduced by using integrated assessment models in order to get an idea of the appropriate tax rate. Finally, there must be an exact correspondence between the tax base and the emission of pollutants, which may not be the case with mobile or diffuse sources of pollutants. In such cases a proxy for the pollutant must be determined.

The dual nature of emission charges and *tradable permits* enables a comparison of impacts from a theoretical point of view. The duality is based on the principle of price fixing. Whereas under emission charges the price is fixed, under a tradable permit system it is variable. The advantage of the latter is therefore, that prices automatically adapt to changes of the economic framework such as economic growth and inflation, so that the total allowed emission is not exceeded (Koutstaal and Nentjes 1995, p. 230). Besides, fixing the total amount of emissions reduces uncertainty about environmental effectiveness and adjustment costs. Without any information about the firm's private marginal costs the control authority can still be sure that the environmental target will be met.

The optimal design and the appropriateness of tradable permits crucially depend on the type of pollutant that has to be controlled. For non uniformly mixed pollutants, such as SO_2, where the pollution target is defined in terms of concentrations or depositions measured at a number of receptor points (e.g. according to the "critical loads" approach[19]), the design of the permit system has to take into account the location of the emissions (Tietenberg 1990, p. 22). Economic theory suggests a wide range of tradable permits models with different spatial dimensions (Klaassen 1996b). The first-best system of tradable permits, in terms of cost-efficiency and ecological effectiveness is that of ambient permits. These permits allow emitters to increase the concentration or deposition of emissions at each specific receptor by one unit. Thus, each emission source has to hold a portfolio of ambient permits for all receptors that are affected by its emissions. The ambient permit solution therefore requires one permit market per receptor point. Unfortunately, ambient permits will fail in practice, because of potentially high transaction costs. Also, the numbers of traders in each market can become rather small, increasing the potential problems from market imperfections. More practicable but less appropriate in terms of costs are the highest ambient permit (HAP) system, developed by Atkinson and Tietenberg (1982), zonal trading models, discussed in Klaassen (1996b), or systems which combine undifferentiated emission trading models with regulatory measures to avoid hot spots (Koschel et al. 1997).

---

[19] The "critical loads" approach has been applied for the first time in the 1994 Protocol to the UN/ECE Convention on further reduction of sulphur emissions. The European Commission, too, has adopted this concept within its Communication "A European Union strategy to combat acidification", presented on March 1997.

A regionally differentiated permit system for SO_2 under a set of emission or deposition standards is a typical example for a trade-off between economic efficiency, ecological effectiveness and administrative practicability. Models with high appeal in terms of ecological and economic effectiveness are characterised by lack of practicability: the higher the ecological goal-conformity, the higher the transaction costs. The parallelism of different permit markets - which is an inherent feature of spatially differentiated schemes - raises the risk costs of emission sources not to obtain all necessary permits and the uncertainty of permit prices. This raises the opportunity costs of relying on permits and thus the attractiveness of technical solutions (e.g. the installation of a flue gas desulphurization process). The loss in welfare from using uniform regulation when spatially differentiated regulation would be appropriate is greater when marginal cost and benefit functions are more steeply sloped (Kolstad 1987).

In the case of locally concentrated pollutants systems of undifferentiated emission permits involve an inherent conflict between ecological effectiveness and economic efficiency. Due to the missing differentiation between the receptors, the risk of hot spots increases. However this type of trading scheme reduces the risk in the long run, as the cost efficiency facilitates the implementation of more ambitious emission reduction plans in future (see below).

The environmental effectiveness of *voluntary agreements* face three potential weaknesses (see Table 3-3):

- an incentive problem;

- an enforcement problem at the level of the industrial association, and

- an enforcement problem at the level of the individual firm.

However, the co-operative approaches risks softening the level of environmental protection, since the environmental targets are the result of a bargaining-process between government and industry. The bargaining power of the government depends on its ability to set up a credible threat with alternative regulation. Hence, the government may have to accept substantive and/or temporal cutbacks in its environmental targets so as to elicit from industry any commitment at all to act voluntarily (Hartkopf and Bohne 1983, p. 229). A consensus often presupposes "a downward correction of goals" (Rennings et al. 1997). However, in the presence of diseconomies of scale in control and enforcement, voluntary agreement could be superior to taxes, as efficiency losses in emission reduction are traded against efficiency gains in control- and enforcement costs (Schmelzer1996).

Goal-conformity will depend decisively on whether the voluntary agreements are legally binding or not binding. If they are non binding, violation of the agreement does not have any legal consequences for the association and such voluntary agreements entail the risk that companies or industry confederations will seek to evade their obligations. Non binding voluntary agreements are thus presumptively goal-conferment only in certain applications, e.g. when the agreed goal triggers only marginal differences in interests between the state and the industry, and does not run counter to corporate self-interest (Murswiek, 1988, p. 988). If more ambitious goals are involved, which require more than no-regret strategies[20], the parallel use of regulatory or economic instruments and/or a credible threat of legislative action in the case of non compliance will be indispensable. Only in this case the government avoids loosing its coercive power. "It can be said that the credible threat defines the space where voluntarism and negotiations arise" (Glachant 1994, p. 41). If environmental agreements are binding there exists a formal sanction mechanism and the incentive problem for compliance on the level of the association is reduced.

There is also an enforcement problem at the level of firms. Since agreements are reached between government and business associations the classical prisoner's dilemma arises if the number of firms of the involved association is high. In order to tackle the resulting free-rider problem, the environmental targets set down in the agreement have to be become operational and disaggregated of the level of the individual firm. In addition, enforcement mechanisms will sanction to those who violate the commitments, have to be introduced.

3.4.2 Static Cost Efficiency

The extent to which regulatory instruments will limit individual options for action depends on their specific nature. In principle, performance standards or emission ceilings, that do not prescribe a certain technology or type of input, offer the individual polluters the same options to react as emission taxes or emission tradable permit system do. Consequently, regulations are not automatically inefficient at the firm level (Klaassen 1996a, p. 26). Only technology-based standards restrict the individual's freedom of choice and raise inefficiency at the firm level. Prohibitions, designed solely to prevent particular ecological damaging actions, are the strictest form of regulation. Nevertheless, they frequently offer a wider degree of freedom than mandatory prescriptive requirements do. Thus, they may comply with free-market principles. If low-cost substitutes for banned products are available, prohibitions may actually be a relatively cost-efficient instrument (Fisher et al. 1996, p. 412).

The main economic criticism of regulations is that they do not permit any allocable efficiency to be implemented at an international, national and sectoral level, when marginal costs differ (Klaassen 1996a, p. 26ff.). This is mainly due to

---

[20] Fisher et al. (1996, p. 411) define no-regret policies as follows: "Policy or policy reforms that improve the efficiency of an economy while also reducing greenhouse gas emissions have sometimes been described as „no-regrets" policies because they offer sufficient benefits in other contexts that their adoption could not be regretted even if climate change were later shown not to be detrimental".

the lack of information on part of governments on the cost functions of individual firms. Uniform technology or emission standards ignore differences in the marginal abatement costs and thus lead to cost inefficient solutions. For "subsequent correction" of inefficient regulations, flexible options, i.e. substitution options for allocated environmental rights of use, are being discussed and/or practised. These, however, basically leave regulatory enforcement untouched, and thus exhibit only slight allocable effects (Hansmeyer and Schneider 1990, p. 59).

However, if the spatial characteristics of pollution are important, as is the case with non uniformly mixed pollutants, the full exploitation of firm- or country-specific cost differences is not desirable as it may lead to violations of the targets, defined in concentrations or depositions. In this connection Tietenberg (1995, pp. 99ff.) has analysed and evaluated empirical studies, which compare the abatement costs of reaching a vector of concentration targets using an emission permit system (which equalises abatement costs completely - see below) with those of a regulation system. He comes to the conclusion that differences in the control costs are due to two components. The equal-marginal-cost component refers to the lower costs of emission reduction associated with the equalisation of marginal abatement costs. The degree-of-required-control component reflects the total amount of emission reduction needed. Since the first component unambiguously favours emission permits, the second component either favours emission permits or regulation, depending on the spatial configuration of sources.

In conclusion, regulations are cost-inefficient if uniformly mixed pollutants, such as greenhouse gases, have to be controlled. But if spatial considerations have to be taken into account the relative cost-efficiency of regulations and economic instruments depends on the specific local circumstances and targets. Imposing a perfectly differentiated tax system (with a location determined tax rate) can become very troublesome, if not impossible.

In comparison to regulatory instruments, under which the legally permissible "residual pollution" remains financially unburdened, *environmental taxes* follow the polluter-pays principle systematically. Environmental taxes provide more scope for behavioural choice, since they constitute an "indirect, behaviour-inducing and not behaviour-prescriptive state intervention" (Reichmann 1994, p. 132). The prospective taxpayer concerned can choose between reducing the use of the environment, or paying the tax involved. The allocable advantages of economic instruments over regulatory instruments are mainly based on the fact that the governmental decision-maker needs no precise information on plant-specific abatement techniques in order to realise a cost-efficient solution. In theory, taxes lead to cost-efficient implementation of environmental goals, since, marginal abatement costs are equalised.

The scope for action provided depends on the type of taxes, which is chosen. Basically two categories of taxes can be distinguished:

- emission taxes, i.e. the taxation of polluting emissions (e.g. carbon tax);

50

- product (excise) taxes, i.e. the taxation of pollution-intensive inputs or outputs (e.g. tax charged on fossil fuels in the form of a carbon or energy tax).

Under an *emission tax* system all polluters face a tax per unit of emissions, i.e. the tax base is the quantity of emissions (Fisher et al. 1996, p. 403). If taxes are primarily used for implementing an emission reduction target, emission taxes are first best in terms of cost-efficiency from a theoretical point of view. Since the tax base is directly related to those emissions, which cause the externality, e.g. climate change, environmental and economic losses will be minimised. Compared to other types of taxes, emission taxes provide the firms with a maximum degree of freedom. In the case of a tax on sulphur emissions, for example, firms can reduce their burden of taxation not only by substituting sulphur-intensive inputs by less intensive ones. Firms also can avoid taxes by using more energy-efficient technologies or they can install end-of-pipe technologies. The implementation of this type of tax implies that the monitoring of emissions is possible and not too costly.

In contrast to emission taxes, *product or excise taxes* are applied to raw materials, intermediate inputs, and to final consumer products. Product taxes are very common in the field of energy. Examples include taxes on gasoline, diesel and heating oils or electricity.

From a perspective of allocable efficiency the direct taxation of emissions is first best. However, because of transaction costs (information, administrative and measurement costs), the implementation of an emission tax often requires the use of alternative tax bases which act as proxies for the actual (not directly observable) emissions (see Figure 3-2). Excise taxes imposed on pollution-intensive products thus may represent second or first-best solutions, if specific efficiency criteria are applied. The arrow in Figure 3-2 symbolises the increase in distortions that arise as alternative emission tax bases are applied. Whereas a tax charged on inputs at least allows input substitution processes on the production level (see below), a tax on pollution-intensive outputs changes relative prices and demand, but does not stimulate any response in abatement. Incentives are even more distorted if the tax is based on capital inputs or sales (Siebert 1992, p. 135).

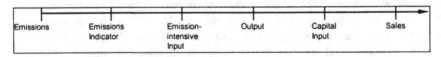

Source: Siebert 1992, p. 137.

Figure 3-2: Possible tax bases of emission taxes.

In the case of CO_2, where emissions are linearly related to the carbon content of fossil fuels and where no end-of-pipe technologies exist, the indirect taxation of emissions via energy inputs is fully equivalent to direct taxation, as far as efficiency is concerned. But this equivalence becomes only approximate if fossil fuels are taxed according to their long-term global warming potential.

When a fixed emission reduction target has to be reached, the taxation of emission-intensive inputs is less efficient than direct emission taxation since the link between taxable base and environmental pollution is less direct. In the case of CO_2, taxes on energy products with rates differentiated according to their externalities coincide with the economic principle of efficiency. In contrast, a tax on energy products based on the energy content such as the US BTU tax in February 1993 (Muller 1996, pp 476f), is only third- or fourth-best in terms of efficiency, since it does not give any desirable incentives for substitution processes between inputs of different carbon intensities (see also Fisher et al. 1996, p. 403).

Excise taxes on fossil fuels, such as coal, natural gas and crude oil, are appropriate to stimulate innovation processes towards increased energy efficiency. Imperfect competition in the resource markets (oil and gas markets) can prevent a fall in primary producer energy prices which could take place I competitive markets and which counteract to one extent the impact of taxes. Energy taxes have positive effect on resource scarcity and natural resource management targets (e.g. protection of scarce mineral resources or energy sources from depletion). Nevertheless, emission reduction targets and natural resource protection goals should not be mixed up, since their efficient design has to be based on different criteria. The criteria of efficiency and effectiveness make different demands on the design of an input tax depending on whether it is used for implementing natural resource management goals or emission reduction targets (OECD 1997, p. 16). A resource tax (e.g. a tax on fossil fuels) can justified if the real market solution differs from the social optimal path of resource use. Such market failures may arise for many reasons, e.g. absence of ideal future markets for scarce resources, differences of interest rates from the appropriate social discount rate or deviations of private from social extraction costs (Endres and Querner 1993, pp. 67ff.). The objective of resource taxation is to slow down the speed of resource extraction.

In March 1997 the EU-Commission proposed a common EU-wide excise duty system which included the taxation of all energy products (i.e. mineral oils, coal, natural gas, that are used as motor fuel or heating fuel) and electricity at the final consumption stage (Europe Environment 1997a). In addition, the Commission proposed obligatory and voluntary exemptions for some energy products as well as for some uses, such as air navigation and renewable energies. The Commission set minimum tax rates which are intended to lead to a closer approximation of national rates in three two year stages, beginning in January 1998, January 2000 and January 2002. Energy value will be the basic criterion for taxation even though the objective of the proposal was environmental protection. The proposal allows the Member States to apply different tax rates, provided these rates are higher than the desired minimum tax level. The degree of cost-efficiency (in terms of global warming) depends on whether the Member States make use of the Commission's suggestion to differentiate their tax rates on energy products according to national environmental goals.

An environmental tradable permit is an instrument that is highly reliant on market mechanisms. The state merely establishes an ecological and permit distribution framework. Trade of permits allows for an efficient allocation of pollution rights, and under some conditions emission permits can realise the least cost solution (Montgomery, 1972). These conditions are that transaction costs are negligible, that the permit market is competitive, and that sources minimise their control costs (i.e. the costs of control technology investments and the net revenue of trading permits). They are less likely to be satisfied in the case of non uniformly mixed pollutants where a separate market for each receptor has to be set up and the numbers of traders may be small. It was also shown by Montgomery (ibid.), that for ambient permit markets, under certain conditions, the initial allocation of the permits has no allocable consequences and thus represents only the distributive element of a permit scheme. The initial allocation represents a lump sum endowment, which does not affect marginal choices, i.e. within a permit system equity and efficiency aspects can be handled separately (Tietenberg 1983, p. 240). This implies that the government or control authority can use the initial allocation of emissions standards to realise distributional goals without interfering with cost-efficiency goals (Tietenberg 1990, p. 22). The absence of the usual equity-efficiency trade-off is an attractive feature of the tradable permits approach, in particular at the international level. For instance, an international CO_2 permit system could address questions of inter-country equity arising from the concept of sustainable development without interfering with the system's efficiency (Rose and Stevens 1996).

Tradable permits grant companies a similar freedom of choice as taxes do, since the individual companies can choose between acquiring tradable permits or reducing their own emissions. The market prices emerging on the markets for permits may be interpreted as "shadow prices for environmental restrictions" (Bonus 1990, p. 350). The design of tradable permit systems differ in terms of how far the spatial aspect of the pollution rights restricts trade in order to avoid deleterious emission concentrations (Rehbinder 1994, pp. 94 ff.). There may thus be a conflict of goals between ecological and economic viability. Thus, tradable permits are especially promising in the core of uniformly mixed pollutants where the origin is of no relevance, or for quantity control of closed systems, such as indirect discharges into the municipal sewage system (Hansmeyer and Schneider, 1990, p. 58).

The efficiency properties of economic instruments, when compared to command and control instruments have been derived under the assumption of perfectly competitive product markets. Sartzetakis (1997) has shown that, in case of imperfectly competitive markets, tradable permits are not necessarily superior to command and control instruments. They achieve cost-minimisation of emission control effort, but through the redistribution of production between firms because of market imperfection and differences in abatement cost between firms, inefficient firms can gain market share. The design of the permit system can however minimise the importance of the output redistribution effect.

Uncertainty regarding the cost and the benefits of an abatement activity is also important for the relative efficiency of quantity instruments (permit system) versus price instrument (tax). Weitzmann (1972), considering the case of cost benefit uncertainty on the part of the regulatory agency, has shown that benefit uncertainty alone does not matter and that cost uncertainty can have a significant impact depending upon the relative slope of the marginal benefit and the marginal cost curves, with a steeper cost curve favouring the price instruments. When benefit and cost are simultaneously present and correlated, benefit uncertainty matters (Stavins 1996). A positive (negative) correlation tends to favour a quantity (price) instrument, the greater the uncertainties and the lesser the slope of the marginal cost function, the greater the impact of the degree of correlation between benefits and costs. Stavins suggests that for uniformly mixed pollutants, assuming no correlation between cost and benefits seems reasonable, but for other types of pollutants a positive correlation is more plausible, making quantity instruments more attractive.

Due to their 'informal' character (no legislation process is required) *voluntary agreements* are more flexible instruments concerning implementation than other instruments. Voluntary agreements often are described as market-based and cost-effectiveness instruments. The degree of market-conformity of voluntary agreements crucially depends on whether the implementation of the agreement's goals and measures at the industrial level uses to market based instruments. The realisation of a cost-efficient allocation of reduction measures presupposes complete information on the part of the government of the firm specific abatement cost or the existence of a uniform price signal for environmental utilisation. Since, voluntary agreements have no systematic tendency to make use of market-based environmental instruments, and in reality do so only rarely, an efficient allocation is by no means assured, and in fact rather improbable. Glachant (1996), however, showed that they can be cost efficient under certain conditions: very large shared uncertainty about pollution abatement technologies, concentrated industrial sectors in which the heterogeneity in pollution abatement activities and costs is low.

3.4.3 Dynamic Efficiency

In terms of dynamic efficiency, *regulatory requirements* are less preferable than economic instruments (Milliman and Prince 1989, Downing and White 1986, Jung et al. 1996). Legislation updates of emission or technological standards are likely to lag changes in the state of the art of technology (Hohmeyer and Koschel 1995).

Regulatory instruments also do not generate incentives for innovations. The residual pollution, i.e. the amount of pollution remaining in the environment after regulation, is costless and no real costs or opportunity costs have to be taken into account. Only suppliers of advanced (end-of-pipe) technologies may have an incentive to develop technologies which go beyond the required reduction level in the hope that they will push ahead the legislative technical standard and to open up new markets for end-of-pipe technologies (Hemmelskamp 1997).

Voluntary agreements do not provide either any economic incentives or stimuli for technological advances going beyond the agreed reduction target (Kohlhaas and Praetorius, 1994, p. 177). Their effect on technical progress is rather low and comparable to those of regulations. However, in case of market failures and imperfect competition or when environmental innovation has positive spillovers on other firms, the use of voluntary agreements can be justified (Carraro and Siniscalco 1996). In strategic industries producing tradable voluntary agreements can be an effective instrument to reduce pollution and to induce environmental innovation (Carraro and Siniscalco 1992).

From a theoretical point of view, the *tax* and the *tradable permit* solution, provide greater dynamic incentive effects for developing avoidance technologies (Hemmelskamp 1997). Both systems give incentives to search for and invest in advanced pollution abatement technologies, but in addition provide incentives for innovation of more cost-effective control technologies. Since unused permits can be sold on the market and the revenue used for financing investment, environmental technical progress will be stimulated permanently as long as any 'opportunity costs' of using a permit exist (Tietenberg 1994). However, tradable permits are frequently described as less dynamically efficient than environmental taxes, since as technical progress advances and is diffused among firms, the total demand for permits will decrease and the permit price will fall (with constant supply). This will result in reduced incentives for further innovations because the opportunity cost of using a permit decreases (Ewers and Brenck 1995, pp. 131 f.). Additional reductions in emissions can (in contrast to the tax solution) be implemented only by boosting the system's dynamics from the outside (Bonus 1994, p. 3), for instance by reducing the total amount of allowed emissions under a permit system in the same way as it would happen automatically under a tax system. In order to minimise the firm's decision and planning uncertainty and to maximise the innovative impact of a permit system, the central authority should pronounce the long-term path of emission reduction in adverse. Under emission charges the cost of polluting is fixed and therefore the incentives for innovation are undiminished during the diffusion process.

Empirical evidence from the U.S. Acid Rain Program (ARP) supports the theoretical finding on the dynamic efficiency of tradable permit systems (Koschel and Stronzik 1996). Currently, the price for a SO_2 allowance (around US\$100/t SO_2) is well below the estimates (around 1.000 US\$/t SO_2) published before the program started. This is partially due to the exogenous changes in fuel and transport markets but also due to the increased flexibility in compliance. The ARP has established competition in input markets and among suppliers of abatement technologies. This has led to numerous innovations in mining (increased productivity) and shipping of coal, as well as in scrubber design and fuel blending.

The impact of tradable permits on innovations depends on the instrument's design, in particular on the initial allocation mechanism, as well as on the instrument's ability to give long-term price signals. Theoretical models show that auctioned permits generate higher incentives to promote technical progress in pollution control than free permits (see Milliman and Prince 1989, Jung et al. 1996). Requate (1995), with a rather general model where attention is extended to the output market (and not only to the pollution sector), shows that neither taxes nor auctioned permits are superior for providing the correct incentives for firms to adopt a new, less polluting technology. Both can lead to non optimal solutions with under or over-investment in the new technology. However, permits allow for adjustment after the innovation and do not reduce welfare, whereas taxes sometimes do.

The policy mix of which the instrument is part of is of importance for its dynamic efficiency. Restrictions on innovative effects may arise if the permit system is merely integrated into an existing command-and-control system. Dynamic technical standards, based on 'state-of-the-art technology', can hinder innovation activities since the emitters may fear that innovations will drive to a general tightening up of technical requirements. Experiences within the U.S. Emissions Trading Program and the German Air Pollution Legislation stress the disadvantages of such 'flexible standards' compared with tradable permits: transaction costs and a close connection to the command-and-control approval system have restricted trade and technical progress decisively.

In this connection the interdependency between the level of environmental targets and policy instruments should be pointed out: cost-efficient instruments such as tradable permits burden the economy with lower costs than an equivalent command-and-control policy. Therefore, the political acceptability of stronger emission targets and incentives for technological progress in pollution control will be higher under market based instruments. These findings are supported by the experiences in U.S. decision-making on targets within the U.S. Acid Rain Program (Endres and Schwarze 1994).

3.4.4 Non Environmental Benefits (Revenue Recycling Issues) and Other Policy Objectives

In contrast to regulatory instruments and freely granted tradable permits, taxes (and auctioned tradable permits as well) lead to revenues which can be used to offset the existing public deficit or can be refunded in order to avoid an increase in the overall tax burden.

A revenue-neutral policy may pursue several goals. Firstly, revenue-neutrality may support distributive goals. For example, tax revenue refund systems for firms with above average shares of energy costs in overall production costs could prevent substantial losses in competitiveness vis-à-vis third countries. Other reimbursements of tax revenues, like a per capita allowance ("eco-bonus") (see Bach et al. 1996, p. 453) are possible. Secondly, revenues can be used in order to create additional environmental incentives. Examples include for instance the public financing of environmentally friendly infrastructure and public services as well as the supporting R&D and investments in energy saving measures (IBID, p. 453). Thirdly, recycling options can be aimed primarily at reducing distortions. Whereas in a first-best world without distortionary taxes a reimbursement of tax revenues via lump sum transfers would be optimal, in a second-best world, following the theory of excess burden and optimal commodity taxation, the tax revenues could be obtained by rationalising the tax system by reducing dead-weight losses[21] from other distortionary taxes such as labour or income taxes.

Frequently, distributive, allocable or fiscal goals arise together, if firms and households as a whole receive a compensation equal to the additional burden from the environmental tax; not as lump-sum transfer but rather in form of reduced wage costs or decreased income taxes. A reduction of employer's social security contributions, as suggested by the Commission in its 1993 white paper 'Growth, Competitiveness, Employment', might also tone down losses in competitiveness. However, energy-intensive manufacturing branches may not be fully compensated. The chance of obtaining additional positive allocable effects in non environmental areas has led to a growing interest in eco-taxes as a tool for climate protection over recent years. In this context the term "double dividend" has been brought into discussion. Whereas the first dividend is related to improvements of the environmental quality, the second dividend generally refers to the economic gains obtained through the reduction in distortionary, pre-existing taxes.

The double dividend issue has taken different forms depending on the definition at the second dividend:

- a weak form of the double dividend, that means, that recycling of tax revenues through cuts in distortionary taxes leads to less costs compared to the case where revenues are returned as lump sums;

- stronger forms of the double dividend, that imply, that the revenue-neutral introduction of environmental taxes, which replace partially or completely taxes that bring about large excess burden, involve zero or negative gross costs (i.e. cost reductions).

---

[21] The dead-weight loss (or excess burden respectively) from a tax system can be defined as that amount that is lost in excess of what the government collects (Auerbach 1985, p. 67).

Certainly, the validity of the strong form of the double dividend facilitate enhance the environment policy-decision making process significantly, since - provided that the environmental benefits are positive - the implementation of an environmental tax reform can be justified on overall benefit-cost grounds. For instance, Conrad (1993) shows that, maximisation of domestic or international welfare in an imperfectly competitive world gives incentives to governments to introduce subsidies for abatement activities or for emission taxed inputs. Obviously, environmental policy might be used to attain non environmental targets. However, looking at optimal government policy in the presence of an externality of the congestion type, and with income distribution taken into account, Mayeres and Proost (1997) arrive at the conclusion that the level of the externality tax does not depend strongly on distribution concerns when other taxes can be adapted to reach the income distribution objective.

Other studies have looked at the impact that international environmental tax competition could have on the level of environmental regulation, assuming either competitive or non competitive markets. Especially with non competitive markets, deviations from jointly optimal policies can be substantial (Hoel 1994, Rauscher 1995).

3.4.5   Administrative Practicability

The question whether economic or regulation instruments are more appropriate in terms of administrative practicability is still not resolved. Tradable permits require less information with respect to marginal costs than taxes or regulations do (Klaassen 1996a, p. 33). The implementation of an environmentally effective tax system requires information on aggregate marginal costs, whereas the regulation approach in order to be both environmental effective and cost efficient needs, in addition, information on source specific costs. However, the administrative practicability of a system of tradable permits is influenced by the type of pollutant being controlled. As aforementioned, in the case of non uniformly mixed pollutants, transaction and information costs might increase significantly. Taxes might be less costly to implement than permits. When there are numerous and relatively small pollution sources, permit transaction costs may be too high.

There is no evidence that monitoring and enforcement costs of economic instruments are higher or lower than those of regulations (Klaassen 1996a, pp. 32ff.). Bohm and Russell (1985, p. 416) estimate that costs of monitoring for compliance are the same for an emission tax, an emission permit system and an emission standard, if all are defined in terms of allowable emissions per unit time. Direct regulations, for instance, are characterised by a system inherent lack of economic incentives for compliance. Therefore, enforcement frequently requires costly legal steps (Hohmeyer and Koschel 1995). Taxes and permits, on the other hand, cause administrative costs that are due to the collecting of the taxes and the supervision of the permit trade.

Klaassen (1996a, p. 32) points out that, in case of exogenous change in economic conditions, the administrative effort to maintain the environmental target constant is highest under a tax system. It is lowest under a permit system, where the price automatically adjusts to these changes. No general conclusion can be drawn, whether administrative costs are higher or lower under direct regulations, taxes or permits. The administrative difficulties of voluntary agreements are related to the potentially large information requirements

3.4.6   Political Feasibility

There are some arguments that the political feasibility of voluntary agreements and direct regulations is higher than that of taxes or permits.

Governmental decision-makers might prefer a policy of direct regulations to other instruments, because of its high visibility and allocability. The resistance to environmental taxes and tradable permits is partly based on a general aversion to institutional innovations and the concomitant reshuffling of prerogatives within civil service departments and corporate hierarchies. On the other hand, taxes can be attractive instruments for politicians if they are refunded, so that other political goals (employment) can be pursued simultaneously.

When implementing command-and-control measures the resistance from industries affected is smaller since regulations increase overall costs less than taxes (but not necessarily less than grandfathered permits and revenue-neutral taxes) and are more calculable for the individual firms than a permit system. Also industries have the impression to have more flexibility in the bargaining and enforcement processes and there are no transfers between firms. The opportunities for political abuse of environmental taxes are seen especially through the possibility that the revenues will be welcomed by the Treasury as an inexhaustible source of income, whereas steering goals of environmental policy will have to take a back seat. This poses a danger to goal-conformity, not to mention the possibility that corporate cost burdens will escalate to an unacceptable level (Bonus 1990, pp. 357f.). Thus the implementation of revenue-neutral tax reform concepts also involves inherent difficulties. However, ensuring revenue-neutrality is more of a political problem than one of taxation methodology.

The introduction of tradable permit systems is, at least in Europe, not high a priority for the time being. This retreat is linked to the lack of practical experience with this instrument and the lack of a consensus regarding the allocable procedures for emission rights. The polluters fear that successes in the reduction of pollutants (which will be directly manifested in a price slump) will encourage the state to immediately enact a more stringent ecological framework. Allocation of permits on a grandfathering basis makes it easier for convincing polluters to join the scheme, but it generates problems with new entry.

3.5 Conclusions

Environmental policy instruments have to be evaluated on the basis of several criteria. Not only must a distinction be drawn between their use for averting dangers and for making provision for risks; the nature of market failure in each case (external costs, inflexibility, lack of information) and the type of pollutant (uniformly dispersed or not, cumulative or not) must also be taken into account. Moreover, any policy has to take into account sectoral specificity. Table 3-4 provides a simplified overview of environmental policy instruments evaluated mainly in terms of cost-efficiency, environmental effectiveness, political feasibility and administrative practicability.

From an economic point of view the use of regulatory instruments can be justified without reservation in situations of averting dangers where individual flexibility of emitters must be sacrificed. In addition, if the pollutant's location matters regulations may lead to control costs that are not higher than those under economic instruments. This will also be the case when monitoring is difficult or costly, as is the case in presence of many small polluters. In all other cases, and in particular in the case of greenhouse gases, regulatory instruments represent rather second- or third-best optimal solutions as they do not realise a cost efficient allocation of abatement efforts (because of information deficits of the control authority) and do not give incentives for the development of new technologies (since the polluter-pays principle is implemented incompletely). They can also be subject to bargaining and negotiations. Nevertheless, because of reasons of administrative practicability and political feasibility they can be preferred, in particular at the level of private households.

Emission taxes and tradable emission permits are theoretically equivalent in terms of static cost-efficiency as they produce identical allocable solutions. In reality, the implementation of tradable permits requires less information than an equivalent tax, since the price is brought out endogenously by the market mechanism and does not have to be calculated on the basis of (uncertain) abatement cost estimations. Therefore, tradable permits are more environmentally effective. But a strong prerequisite for a cost-efficient allocation under a permit system is the existence of fully functioning markets with price signals reflecting the (politically determined) scarcity of goods correctly. Market distortions caused by market power or strategic behaviour therefore may endanger the efficiency of the market's allocation. Both, taxes and permits are dynamically efficient as they generate an economic self-interest in developing advanced abatement technologies. Whereas under a permit system technological progress does not lead automatically to reductions in the overall emissions, but finds expression in a drop of the permit price, under an emission tax technical progress will be directly translated into emission reductions. The tax base must be related to, as closely as possible, the environmental target.

Recently, voluntary agreements have become more significant in political discussion and practice. Even if voluntary agreements have special advantages due to their co-operative approach, they present many difficulties. Unquestionably, they are most successful if they are binding and provided with a monitoring and enforcement system. Voluntary agreements can be used alternatively or additionally to other "tough" environmental instruments such as environmental taxes or regulatory instruments, if they are used alone they are unlikely to keep with market developments and do not normally send out correct price signals Cost-efficiency can be realised if voluntary agreements are used in combination with economic instruments. Voluntary agreements may facilitate the transition to a polluter-pay oriented environmental policy (COM (96) 561 final, p. 8). For instance, voluntary agreements could be used in order to spare sectors from high cost disadvantages in international or national competition by exempting them from taxation during the transitory period. Nevertheless, a more appropriate option for minimising undesirable adjustment costs is likely to be a gradualist policy, in the form of taxation, which starts with low taxes that increase over time. The use of an effective instrument, such as taxes or permits, from the beginning avoids the danger of intervention cycles. In order to ensure the achievement of the targets set by environmental policy the government may be forced to launch a cascade of new policy instruments. The consequences of such a policy might be increasing regulatory intensity and administrative costs as well as decreasing transparency of the regulatory framework (Rennings et al. 1997).

Table 3-4: Evaluation of environmental policy instruments.

	Environmental effectiveness				Economic efficiency		Non environmental benefits	Political feasibility	Administrative practicability
	Averting dangers	Provision against potential risks			Static	Dynamic			
		Information	Lack of information	Supplementary function					
Regulatory instruments	+	+	-	-	-	-	-	+	+
Emission/excise tax	-	+	-	-	+	+	+	-	+
Tradable permits	-	+	-	-	+	+	+/-	-	+
Voluntary agreements	-	+/-	+/-	+/-	-	-	-	+	-
Liability law	-	-	+	+	+	+	-	+	+
Subsidies	-	-	-	+	-	-	-	+	+

+ = rather high; - = rather low; +/- = depending on circumstances

4 Baseline Scenario for the EU and Overview of Policy Simulations

4.1 Introduction

Following the theoretical presentation of the previous chapters, a large number of simulations were carried out with the latest version of the GEM-E3 model. The main goal was to investigate the impact of reaching alternative QELROS (Quantitative Emission Limitation and Reduction Objectives) for the European Union by the year 2010. The evaluation of the impact of the environmental policy was then derived by comparing the results of the policy runs to a baseline scenario, that was again derived with GEM-E3 but under the assumption that no new environmental policy would be implemented in the EU till 2010.

The model simulations were then used to construct a marginal cost curve of emission abatement. The environmental goal was imposed as a constraint in the economy. The burden of reaching the desired target, was implemented in two alternative ways: as a tax paid by the polluters (the level of the tax was endogenously decided to allow the specified yearly target to be reached), or in the form of tradable pollution permits (whose equilibrium price was again decided by the model).

In the case of tax simulations a number of revenue recycling schemes were also examined (such as reducing the labour costs, or subsidising investment). In the case of pollution permits simulation, the impact of different initial allocation schemes for the permits for burden sharing was also examined.

A modified model version was used to identify the effect of assuming the existence of potential for productivity gains in energy use. This model version identified an energy-saving possibility in production/consumption and gave the possibility to economic agents (as part of their optimal allocation of resources) to invest endogenously in energy-saving technology thereby reducing their energy bill. In this way, the macro-economic model simulates the considerations that prevailed in the analysis of Policies and Measures that accompanied the Council's decision of March 1997.

Another model simulation consisted of an attempt to formally internalise environmental externalities. The valuation of damages (as computed by the EXTERNE project and used in GEM-E3) were incorporated in the decision of the economic agents.

A large number of sensitivity simulations complement the analysis. These simulations concern a number of issues: burden sharing among countries, the impact of exempting some industrial sectors, alternative assumptions on the behaviour of the rest of the world and so on.

4.2   The Baseline Scenario: the European Economy to 2010

The baseline scenario[22] simulates a dynamic path of the EU economy up to 2030. It is derived from exogenous assumptions about the evolution of technology progress associated to production factors, the change of the world context (prices and demand) and the continuation of current patterns of public finance policy. In the present project however, the projections are used until 2010. The full results of the baseline scenario are presented in the appendix.

4.2.1   Introduction

Deriving projections for the longer term is very difficult given the considerable uncertainties involved.

- One difficulty is projecting aggregate regional GDP and population growth. Several studies are available concerning the medium and long term (usually up to 2020) covering the whole planet such as with the LBS-EGEM model (which however does not include Europe) or the world projections prepared by the OECD (LINKAGE project), where Europe is just one region. This latter study was the starting point for the present projection.

- A second, and perhaps even more elusive, difficulty reflects the sectoral changes. As the past has shown these changes can often be very dramatic and move in different directions in different countries. No study is available at the

---

[22] The baseline scenario is the one previously prepared for the European Commission and reported in "European Macro-economic Projections for Baseline Scenario", March 1997, Report of NTUA to the EC.

level of disaggregation needed for energy models, so this part of the projection after 2000 is original.

- The present scenario draws from the available macro-economic and sectoral projections for the short term (up to 2000) and then uses the aggregate world assumptions of the LINKAGE model, extending them to 2030. From a sectoral perspective, there has been an attempt to build a separate "story" describing the evolution of each EU country. The projections were made in two steps:
    - the GDP of each country was derived by assuming gradual conditional convergence of the EU economies in terms of per capita income;
    - on the basis of the present situation of each country and through identifying already existing trends, the driving forces of growth for each economy were identified and served to derive sectoral growth.

4.2.2   Short Run Projections: 1995-2000

The main sources consulted for the preparation of the projections for the period 1996-2000, included the following.

The *DGII projections* for each member state, from which GDP, private consumption, consumers' price indexes, GDP deflator, interest rate and exchange rate were taken.

Sectoral projections were taken from the *DRI study "Europe in 1999 - Economic analysis and Forecasts"*. These projections where only available for 6 EU member states and covered only some of the industrial sectors. Information on services was derived from the *HERMES projections* (again for six countries), which were also used to cross-evaluate the DRI assumptions. Both studies only gave one average figure for the whole 1996-2000 period, so assumptions had to be made about the changes in each particular year. Usually, the assumption of a linear trend was followed. For the other EU countries no sectoral forecasts where available, so current trends were assumed to continue, in such a way however, that the EU-total for each sector matched the DRI forecasts.

Table 4-1: GDP projection up to 2000.

Short run GDP projections (annual growth rates)						
	1995	1996	1997	1998	1999	2000
Austria	2.81%	1.00%	1.60%	2.50%	2.90%	3.10%
Belgium	0.90%	1.40%	2.19%	2.66%	2.90%	3.04%
Germany	1.90%	1.41%	2.16%	2.80%	3.02%	3.21%
Denmark	2.76%	2.08%	3.06%	2.96%	2.86%	3.11%
Finland	4.21%	2.13%	3.67%	3.11%	3.39%	3.56%
France	2.14%	1.09%	2.09%	2.66%	2.85%	3.27%
Greece	2.00%	2.37%	2.51%	2.78%	3.12%	3.51%
Ireland	9.84%	7.75%	5.81%	5.22%	4.79%	4.60%
Italy	2.96%	0.84%	1.44%	2.57%	2.74%	3.03%
The Netherlands	2.08%	2.45%	2.77%	2.95%	3.06%	3.07%
Portugal	2.18%	2.51%	2.81%	3.13%	3.48%	3.68%
Spain	2.85%	2.10%	2.67%	3.17%	3.66%	3.71%
Sweden	3.01%	1.72%	2.11%	2.49%	2.78%	2.96%
UK	2.38%	2.31%	3.00%	3.00%	2.94%	2.99%
EU Average	2.4%	1.6%	2.3%	2.8%	3.0%	3.2%

4.2.3   Long Run Projections: 2001-2010

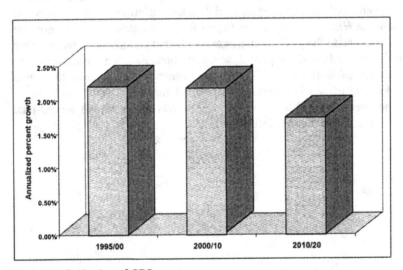

Figure 4-1: Projection of GDP.

The assumption about average EU growth is somewhere in-between the two variants of the latest *World projections*, performed with the *LINKAGE* model for the OECD. In the baseline scenario average EU growth is assumed to increase by 2.4% in the decade 2001-2010, slowing the following decade. The EU population is assumed to increase slightly until 2010 due to immigration.

Table 4-2: Percent difference from EU average in per capita GDP.

	2000	2030
Austria	9.31%	3.60%
Belgium	6.23%	3.00%
Germany	4.35%	3.70%
Denmark	35.35%	25.00%
Finland	26.19%	13.00%
France	11.85%	3.70%
Greece	-67.23%	-50.00%
Ireland	-0.22%	7.00%
Italy	11.94%	7.00%
Netherlands	3.29%	3.60%
Portugal	-57.96%	-38.00%
Spain	-28.17%	-16.00%
Sweden	28.54%	20.00%
UK	-1.71%	1.38%

The GDP projection was then elaborated by member-state following an assumption of gradual convergence of per-capita incomes within the EU. Even in 2010 however, this convergence is far from complete. Based on the projections of changes of inter-EU differences in per capita income, the general EU GDP growth and demographic assumptions, the implied GDP growth per country was obtained. The assumption of convergence implies higher growth rates for the cohesion countries (Greece, Ireland, Portugal, Spain), growth above EU average for the UK and lower growth levels for the rich Scandinavian countries (Denmark, Sweden and Finland). This trend is already present to some extent as can be seen from the high growth rate experienced by Ireland, Spain and Portugal in the period 1985-95. In these comparisons across EU member states, market exchange rates have been used (relative to the ECU), rather than purchasing power parity indicators.

By keeping constant the average EU growth as given in Figure 4-1 and applying the convergence process given in Table 4-1, GDP growth rates by country were computed in a consistent manner. These are shown in Figure 4-2 and Table 4-3 bellow.

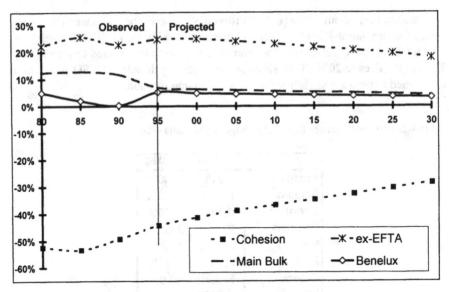

Figure 4-2: Evolution of growth rates in the EU 1980-2030.

Table 4-3: Annualised percent change in GDP.

	Observed			Forecast	
	1980-85	1985-90	1990-95	2000-05	2005-10
Austria	1.31%	3.22%	2.28%	2.19%	2.14%
Belgium	1.02%	2.85%	1.51%	2.29%	2.22%
Germany	1.23%	3.23%	1.84%	2.48%	2.30%
Denmark	2.19%	1.82%	1.68%	2.19%	2.09%
Finland	2.85%	3.39%	-0.69%	2.26%	2.13%
France	1.42%	2.99%	0.94%	2.30%	2.20%
Greece	1.49%	1.68%	1.40%	3.41%	3.72%
Ireland	3.33%	5.57%	6.30%	3.79%	2.67%
Italy	1.42%	3.00%	1.33%	2.12%	2.07%
Netherlands	1.28%	3.06%	1.82%	2.64%	2.54%
Portugal	0.82%	8.04%	1.72%	3.45%	3.62%
Spain	1.46%	4.45%	1.51%	2.91%	2.83%
Sweden	2.24%	2.80%	0.92%	2.30%	2.14%
UK	3.47%	3.25%	1.27%	2.59%	2.54%
EU Average	1.73%	3.22%	1.43%	2.45%	2.35%

Fiscal and Monetary Policy

It is assumed that monetary unification around 2000-2005 will tend to eliminate fluctuations in interest and exchange rates and lead to a gradual convergence of prices. Additionally, a combination of tight monetary policy and the high intra-EU level of competition is assumed to keep price increases and inflation below the 1995 rate.

This process implies that growth of private consumption will be lower than average GDP, leaving more room for financing investments, in the medium run. Gradually a progressive shift is projected leading, after 2020, to equality of growth rates in consumption and investment.

Tight fiscal policies are assumed to prevail over the next decade aiming to reduce public deficits. After that time, a shift in the role of government, towards regulation instead of provision, will result in an increase of the public sector at a rate lower than GDP.

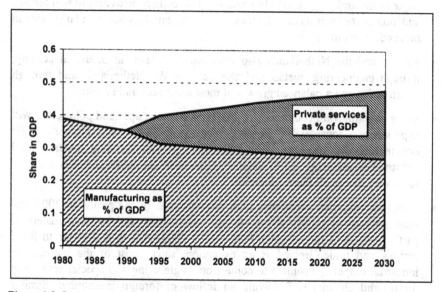

Figure 4-3: Long run sectoral projection.

From a sectoral point of view (Figure 4-3), the baseline assumes a continuation of current trends. Specialisation of countries occurs but in most cases it is not dramatic. Services increase their dominance but do not encompass the whole economy. Industrial activity is rather stabilised following a period of re-structuring. New industrial activities with high value added and lower material base are projected to emerge in most countries. Considerable differences are assumed in sectoral evolution among EU countries. Some of the main assumptions of the scenario are as follows.

- In Germany and France industrial growth is assumed to occur mainly through engineering, chemicals and the non ferrous sectors (to a lesser degree).

Building materials, paper and pulp and non metallic minerals will also increase but less than GDP. Production will stagnate in construction, agriculture, textiles and the iron and steel sectors. The growth of services will continue to be higher than that of GDP.

- Similar trends are projected in the UK. Activity in the manufacturing sector is maintained close to the 1990 levels, mainly through chemicals, equipment and paper and pulp. In Italy, a general slowdown of industrial activity is projected, except in building materials and equipment goods. Activity in textiles and food processing is preserved. Increase in services is projected not to be very high when compared to the EU average.

- In Belgium, manufacturing increases more than GDP up to 2005 and keeps some of its momentum thereafter. Growth comes mainly from chemicals, engineering and construction but also from specialised activities in non ferrous and non metallic minerals. Services increase, but less spectacularly than in most other countries.

- Austria and the Netherlands also maintain their level at industrial activity, through engineering, metals and chemicals in the Netherlands and through construction and a balanced growth of most other sectors in Austria.

- In the Scandinavian countries, it is assumed that the paper and pulp sector will experiences a considerable growth (especially in Sweden and Finland), while engineering will increase faster than GDP growth. Building materials and construction are projected to stagnate. Growth in the public sector is assumed to be very low in all these countries.

- Cohesion countries experience high growth rates. In the period to 2005, the main sources of industrial growth are construction and building materials, partly because of the cohesion funds. After 2005, growth is diverted to food, textiles and trade. Ireland is an exception, since most of the increase in industrial output is assumed to come from engineering and special activity in metals and chemicals, following an inflow of foreign investment. Similar trends are also projected to occur in Portugal. The increase in the share of market services in all these countries is assumed to be above 4% until 2010 and falls gradually thereafter.

The main assumptions of the projections from a sectoral viewpoint are summarised in the following table.

Table 4-4: Main sectoral assumptions of baseline scenario.

Sector	Assumptions
Iron and Steel	Following the decreases observed between 1990-95, production continues to fall undergoing extensive restructuring, to achieve stability the next decades.
Non ferrous	Following considerable restructuring between 1990-95, the sector rebounds achieving stabilisation and growth. This effect is present in Germany, the Netherlands, Spain and the UK.
Chemicals	Production of basic chemicals is projected to be maintained in some countries and especially in France, Germany and the Netherlands. The higher value-added brands of the sector (such as pharmaceuticals) are assumed to flourish benefiting from an increase both in domestic demand and in exports. Growth rates similar to GDP are experienced by Germany, France, Ireland, Italy, Netherlands and Belgium. In the southern cohesion countries, on the other hand, production is assumed to stagnate or increase only marginally.
Paper and pulp	The sector is assumed to face a significant increase in demand especially from the rest of the world. The EU is assumed to continue to be a net exporter. Growth in this sector is assumed to be one of the main driving forces of economic growth in the Scandinavian countries.
Non metallic minerals	Activity in traditional building materials is projected to increase especially in the South, benefiting from an increasing demand from the construction sector, especially in the first decade of the next millennium. In the North activity will be mostly diverted to specialised products with high value added (such as high quality ceramics etc.).
Engineering	After a decade of relative stagnation and restructuring, the sector is assumed to strongly rebound being one of the main driving engines of economic growth for the most important European economies: Growth above GDP is projected for Germany, France, UK, Ireland and Italy. Significant increases are also projected for Belgium and the Netherlands.
Construction	The outlook for construction appears better in the countries of the south where it will be significantly higher than GDP, also because of the cohesion funds. In Portugal and Greece for example the increase in construction in the decade 1995-2005 is of the order of 4% and only slightly lower in Spain.
Textiles	After almost a decade of considerable restructuring that lasts up to 2000, the sector is assumed to rebound in the southern countries: Greece, Italy, Portugal and Spain all face a growth in their production between 2000 and 2010. In the other countries production is assumed to stabilise in level lower than 1990.

Table 4-4: Continued.

Food and Agriculture	Some growth is expected but less than GDP. These sectors are generally assumed to undergo significant restructuring. Higher increases are projected in the south (especially in Greece and Portugal where after 2005 these sectors are expected to start growing significantly again) and to some extent in France.
Energy	The growth in the energy sector is expected to be of the order of 1-2% for most countries, growing at about half the GDP rate.
Services	The role of services is assumed to increase significantly in all countries. Their growth rate is everywhere higher than average GDP. Their share in the economy increases over time but at a decreasing speed. Still increase in activity in private services remains more than 2% per annum until 2010. The increase in the service sectors (especially tourism) is more pronounced in the South.

4.3   Environmental Constraints in the GEM-E3 Model: an Overview of the Model Mechanisms

Within the *GEM-E3, an emission reduction target is imposed as a constraint in the model.* In all the simulations conducted for the present volume, this constraint is *applied to the EU as a whole,* imposing that total CO_2 emissions EU-wide have to be reduced by an exogenously specified amount with respect to the previous year.

Since the constraint refers to a reduction of emissions, in the optimal solution of the model, it will be binding. The shadow cost of this constraint is the marginal abatement cost, which the economic agents bear. Due to the fact that no emission abatement target is imposed by agent category, the economy will allocate the effort where it is cheaper to do so.

The allocation of the cost of emission reduction among economic agents induces an income re-distribution mechanism that affects all EU economies. If the initial situation of the economy were assumed Pareto optimal, such a mechanism alone, would reduce the available resources and thus would necessarily lead to adverse impacts on consumer welfare. If on the other hand, it is assumed that the initial situation from which the EU economies start, involves distortions (such as for example distortionary taxes, unexploited technological potential, terms of trade potential etc.), then the income redistribution resulting from environmental policy can be directed towards reducing these distortions thereby alleviating or inverting the negative impacts. These re-distribution policies define the *"accompanying measures"* of the environmental policy.

Whether the emission reduction is accomplished by means of a tax or a permits market is equivalent in a CGE model like GEM-E3, since markets are assumed to be fully functioning and agents operate with full information, the outcome under a tax or a permit system (under the same assumptions) is theoretically equivalent even if the mechanism through which they operate is different: taxes directly increase the cost of energy and induce higher production/user costs and substitution in production and consumption. Permits, do not, in a first approximation alter relative prices. If however, permits can be traded, then the total cost of production/consumption is reduced by the amount of permits sold. So it is again profitable to substitute energy up to the point where the marginal cost of substituting one additional unit of energy is equal to the permit price.

Regarding the accompanying measures, two main paths can be identified:

- the government intervenes by collecting the environmental tax burden, or by auctioning the pollution permits. An additional issue is how the resources accumulated by the government are re-distributed;

- the pollution permits are directly allocated to the economic agents according to some specified principle (e.g. "grandfathering", or any other scheme) who are then allowed to trade them. Government does not intervene in this option.

In the pollution permits case, agents are given the possibility to sell the emission rights that they do not need or to buy more if their marginal cost of abating is higher than the price of pollution permits (which is equal to the uniform shadow cost of the environmental constraint).

Imposing an environmental constraint in the model gives rise to the following mechanisms:

- a uniform price is assigned per unit of CO_2 emissions. At the optimum, this price is equal to the marginal cost of abatement in all sectors and in for all agents

- the relative cost of energy increases (either directly in the case of a tax or auctioned permits, or indirectly in the case of grandfathered permits)

- in the absence of any other policy, production/consumption costs increase. Substitutions away from energy and towards other production factors/consumption goods are effected. Emission reduction by country and agent is allocated so as to minimise total costs

- in the absence of accompanying policies, competitiveness deteriorates[23], domestic demand decreases, and consumer welfare and GDP are both negatively affected. However the switch from a mainly imported good

---

[23] It is assumed that the environmental goal is applied only in the EU so that prices in the rest of the world remain unaffected.

(energy), to domestically produced ones (capital, labour, materials) compensates competitiveness losses, so that the effect on trade is uncertain

- in the model version including energy-saving investments, the increase of energy costs will augment the profitability of acquiring energy-saving technology. The energy saving stock expands, increasing the productivity of energy in the longer term

- the type of income re-distribution and the secondary effects on the economy depend on the initial allocation (in the case of permits) or on tax revenue recycling (in the case of a tax)

- if there is scope for revenue recycling, some other tax may be reduced, or the revenues can be distributed according to some rule:

  - if they are distributed in order to reduce labour costs, substitution in production is mainly directed towards labour, increasing labour demand (this corresponds to the double dividend hypothesis). In this case, an upward sloping labour supply curve (as in the GEM-E3 model) will lead to a real wage increase. Terms-of-trade gains with respect to the rest-of-the-world, together with an increase in domestic demand (from private consumption) will alleviate part of the competitiveness losses

  - another way of recycling the environmental revenues is to use them to subsidise investments. In this context the cost of investment is lowered so that economic agents are given additional incentives to invest. The overall effect on demand is negative.

4.4 Definition of the Policy Scenarios

Several scenarios were constructed to quantify the macro-economic cost of achieving alternative emission reduction targets.

1. *Construction of a cost-abatement curve for the EU*, for different emission reduction levels (-5% to -25% from the baseline in 2010) in the presence of a tradable pollution permits market. In this, a single goal is applied for the EU as a whole. The pollution permits are grandfathered. Their presence is expected to allow for maximum flexibility in the allocation of emission reduction effort as the agents can adjust through trade of emission rights. This simulation is the reference emission reduction scenario, to which all others are compared.

2. *The role of accompanying policies.* In these simulations, identical emission targets are imposed as above, and this shadow price of the CO_2 emission constraint is considered as an additional cost to all economic agents applied

uniformly and proportional to their emissions. It is further assumed that public finance collect the counterpart of all these additional costs. The underlying assumption is that public policy can then be used to redirect this amount resources to reduce distortionary taxes that may be present in the economy. In this way, it can be expected that part of the negative implications of the environmental constraint can be alleviated, leading to potential economic dividends apart from the environmental ones. Two such cases are considered:

- case of recycling through reducing labour cost. This corresponds to the so-called double dividend analysis, expecting that benefits will arise both for the environment and for employment;

- case of recycling through relaxing financial constraints to investment. The rationale behind this policy is that, by giving incentives to investments, part of the negative implications of the environmental policy can be reversed in the longer term.

A final simulation involves no recycling of the environmental costs. This assumption will obviously have the most negative implications of all the scenarios studied, as the economic agents receive no compensation for the emission reduction burden.

3. *Burden Sharing*. The simulations above derive the EU-optimal allocation of effort and no consideration has been given to the differential effects on member states. In other words, the above simulations. However, equity or other considerations may imply a different allocation of effort. Such an analysis is carried out with the use of pollution permits. The initial endowment is allocated in different ways so as to make sure, for example that the burden per capita is equal, or that the burden per unit of GDP is the same. This allocation obviously implies that countries will have to devote different efforts from those in simulation 1.

4. *Internalisation of external costs*. Instead of imposing a target on emissions, this simulation assumes that agents take into account in their decision process the full cost of using fuels (i.e. the real cost augmented by the "external" damage caused by the emission of pollutants). The GEM-E3 model computes, through the EXTERNE data, the damages caused by the concentration of atmospheric pollutants.

5. *The impact of energy saving technology*. In the standard version of the GEM-E3 model, investments only affect the stock of productive capital and substitution of production factors (through their relative costs) are the only way to reduce the share of some inputs in production, as technical progress is exogenous. This scenario employs a modified version of the model, in which economic agents can specifically invest in the acquisition of energy-saving capital. The amount of investment is not ad-hoc, but rather stems from the

inter-temporal optimal allocation of resources of the economic agents. The link between energy-saving stock and productivity of energy in production is provided by engineering estimates.

6. *Sensitivity tests.* The results obtained in the previous scenarios are naturally sensitive to the modelling and data assumptions made in the model.

In all the above simulations a number of assumptions remain active:

- the equilibrium is obtained simultaneously in all the member-states linked through endogenous bilateral trade flows, formulated according to the Armington assumption;

- trade of the EU with the rest-of-the-world can freely adjust, while the price setting behaviour of the rest-of-the-world remains invariant;

- producers and consumers choose the optimal allocation of production factors (all inputs adjust flexibly) restricted by a production possibility frontier or a pattern for budget allocation of consumers;

- unemployment, exogenously given, is assumed in the baseline scenario, allowing a flexibility of labour force supply. However, real wage rates can vary responding to changes of labour demand, so as to represent the bargaining power of labour unions.

The model evaluates emissions, concentrations and damages. The latter are quantified via the EXTERNE coefficients. Welfare is measured by the model through the concept of *equivalent variation*. Welfare is evaluated at two levels: economic welfare that excludes environmental externalities but includes gains from higher employment; total welfare that also includes the value of avoided environmental damages (from all pollutants, not only from CO_2).

The emission reduction scenarios assume a gradual imposition of the constraint, starting from 2001. The dynamic runs are completed in 2010 when the full target is imposed.

The scenarios conducted with the model are summarised in the following table.

Table 4-5: Main scenarios carried out with the GEM-E3 model.

Simulations	Emission reduction range	Accompanying policy	Sensitivity tests
Pollution permits (PP)	-5% - 25% from baseline	Not applicable	Several, concerning initial allocation of emission rights and burden sharing
Emission constraints	-5% - 25% from baseline	Reduction of labour costs (DD) Investment subsidisation (CR)	Several, concerning sensitivity of labour market, world context
Energy saving investments (ES)	-5% - 25% from baseline	Depends on assumption about the imposition of the goal	In the presence of pollution permits or emission constraints
Internalisation of externalities	Total costs (including environmental damage) taken into account by the economic agents. Emissions computed as a result	No	No

5 The Cost of Meeting Emission Reduction Targets: Pollution Permits

A market for pollution permits is created when a limited amount of "property rights" on emissions are distributed to economic agents. These represent a "right to pollute", proportional to the amount of property rights owned by the agent. These rights can be traded between economic agents (and between EU countries).

The rights are distributed according to a grandfathering[24] principle. An economic agent then has to compare the cost of reducing emissions below its endowment, to the benefit from selling his permits to the market. At the equilibrium point, the permit price will be equal to the marginal cost of abatement.

The pollution permits case (*PP*) does not assume any accompanying macro-economic policy that would aim at removing some other distortion and at obtaining additional gains from that removal. Five different scenarios aiming at emission reduction of CO_2 ranging from 5%-25% have been constructed given the names PP5-PP25. A co-ordinated policy has been assumed, where the permits are traded between all European sectors and households in order to realise the percentage reduction of total CO_2 emissions.

---

[24] A theoretical description of pollution permits is provided in a previous chapter.

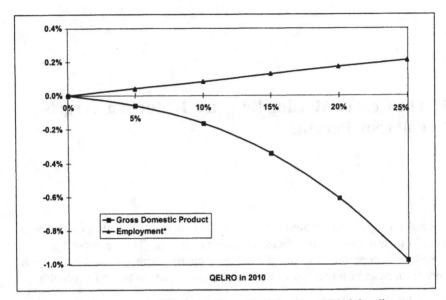

Figure 5-1: Employment and GDP for the European Union in year 2010 for all cases.

In the event of an environmental or energy tax a direct increase in the cost of energy is observed. This results, through the substitution effect, in higher production costs. In the case of permits, on the other hand, the effects are not so obvious since permits do not alter the relative prices. However, they have the same effect with the tax since the producer, under the profit maximisation assumption, prefers to substitute energy (if it is less expensive) in order to sell the permits that he owns. In Figure 5-1 it is obvious that the gross domestic production exhibits a reduction after the imposition of the environmental constraint while employment increases.

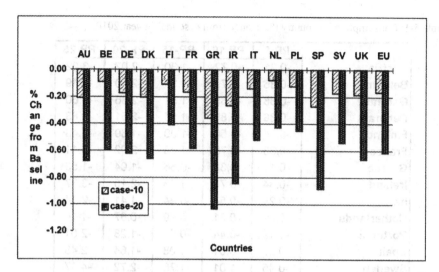

Figure 5-2: Impact on GDP of a 10% and 20% reduction of CO_2 emissions for the EU[25]

All countries experience a reduction in their domestic production, with Greece being the country that experiences the highest decline. In Figure 5-2 it can be seen that other countries that experience high reduction on GDP include Spain and Ireland. Greece's gross domestic product drops by more than -1% for PP20 in year 2010 while in the case of Spain and Ireland a reduction of -0.85% and -0.88% respectively, is observed for the same scenario. The country that shows the lower reduction is Finland with -0.10% in GDP for PP10 and -0.41% for PP20.

---

[25] Numbers indicate percent differences from the baseline and refer to the 2010, the year in which the target is fully achieved. The figures for employment represent absolute difference, while environmental damages are included in the computation of the equivalent variation.

Table 5-1: Consumption by country (% change from baseline) in year 2010.

	PP-5	PP-10	PP-15	PP-20	PP-25
Austria	-0.31	-0.72	-1.30	-2.07	-3.22
Belgium	-0.36	-0.79	-1.37	-2.10	-3.19
Germany	-0.38	-0.85	-1.51	-2.36	-3.60
Denmark	-0.38	-0.87	-1.55	-2.51	-3.83
Finland	-0.25	-0.56	-1.00	-1.60	-2.46
France	-0.33	-0.75	-1.33	-2.02	-3.11
Greece	-0.16	-0.38	-0.66	-1.04	-1.53
Ireland	-0.34	-0.77	-1.38	-2.15	-3.37
Italy	-0.24	-0.54	-0.94	-1.41	-2.14
Netherlands	-0.09	-0.21	-0.39	-0.67	-1.16
Portugal	-0.19	-0.44	-0.81	-1.28	-2.01
Spain	-0.27	-0.61	-1.08	-1.64	-2.48
Sweden	-0.45	-1.01	-1.78	-2.72	-4.17
Un. Kingdom	-0.24	-0.59	-1.12	-2.02	-3.10

In all countries private consumption is lower under a CO_2 constraint. This can be explained partly by the reduction in GDP, and partly by the fact that the durable goods become more expensive due to their energy intensity so consumers shift away to other less expensive products by reducing their consumption. Table 5-1 shows that Netherlands is the country that experiences the least decrease in the private consumption. Greece would be expected to present a high decrease in consumption since it is the country with the higher decrease in the GDP; however, the reduction in Greece's private consumption is found to be very moderate.

Table 5-2: Volume of imports by country (% change from baseline) in year 2010.

	PP-5	PP-10	PP-15	PP-20	PP-25
Austria	-0.48	-0.98	-1.57	-2.26	-3.08
Belgium	-0.27	-0.60	-1.01	-1.61	-2.23
Germany	-0.39	-0.85	-1.44	-2.20	-3.13
Denmark	-0.64	-1.29	-2.04	-2.88	-3.89
Finland	-0.39	-0.76	-1.17	-1.63	-2.16
France	-0.54	-1.15	-1.88	-2.71	-3.78
Greece	-0.45	-0.97	-1.58	-2.33	-3.12
Ireland	-0.45	-0.89	-1.41	-1.97	-2.72
Italy	-0.72	-1.45	-2.29	-3.19	-4.31
Netherlands	-0.24	-0.45	-0.68	-1.04	-1.39
Portugal	-0.31	-0.68	-1.14	-1.69	-2.40
Spain	-0.70	-1.50	-2.47	-3.57	-4.93
Sweden	-0.59	-1.22	-1.96	-2.76	-3.81
Un. Kingdom	-0.29	-0.63	-1.06	-1.69	-2.35

As was mentioned above, a permit policy brings about an increase in production costs and in the prices for domestically produced goods. As a result, domestic demand, and exports are reduced. This reduction, together with the decline in exports, is responsible for the overall production level in EU-14. However, the quantity of imports is reduced, as the substitution effect from domestic to foreign goods is not high enough to compensate the negative income effect. As it is shown in Table 5-2, the total volume of imports declines in all countries. Belgium is the country that experiences the lowest decrease in the volume of imports, -0.60% for PP10 and -1.61% for PP20 in year 2010, whereas Spain reduces its imports by -1.50% and -3.57% for the same cases and year respectively.

Table 5-3: Volume of exports by country (% change from baseline) in year 2010.

	PP-5	PP-10	PP-15	PP-20	PP-25
Austria	-0.11	-0.24	-0.36	-0.45	-0.45
Belgium	0.04	-0.02	-0.15	-0.46	-0.67
Germany	0.16	0.34	0.52	0.71	1.06
Denmark	-0.22	-0.48	-0.71	-0.82	-1.01
Finland	-0.09	-0.19	-0.29	-0.34	-0.41
France	0.10	0.20	0.31	0.34	0.63
Greece	-0.49	-1.25	-2.21	-3.26	-4.79
Ireland	-0.19	-0.41	-0.65	-0.91	-1.17
Italy	-0.14	-0.32	-0.54	-0.87	-1.17
Netherlands	-0.16	-0.38	-0.65	-1.07	-1.57
Portugal	0.03	0.00	-0.05	-0.15	-0.26
Spain	-0.19	-0.53	-1.01	-1.69	-2.46
Sweden	-0.05	-0.10	-0.12	-0.09	0.04
Un. Kingdom	-0.12	-0.28	-0.44	-0.34	-0.51

The EU-wide pressure of costs leads 2010 exports to fall by -0.17% for PP10 and by -0.47% for PP20. For all countries imports decline by a higher percentage than exports. This will have a positive net effect on the balance of trade. A lower import volume for the EU implies a lower export volume for the rest of the world. As it is demonstrated in Table 5-3, Greece is the country that presents the higher reduction in its exports (-1.25% for PP10 in year 2010 and -4.79% for PP20 for the same year). This fact can explain the size of GDP decline of Greece, since the domestic demand and the volume of exports form the production level of the country.

Although the overall level of exports declines, the sectoral patterns differ. In particular, positive growth rates are obtained for exports of fossil fuels, the services and the equipment goods industries. Especially for these sectors, the decrease in domestic production lowers their prices and makes these goods more attractive to the rest of world.

The EU-wide introduction of a permit system leads to an increase in production costs, in particular in energy-intensive sectors, which produce above-average CO_2 emissions. Due to this increase, substitution processes from energy-intensive sectors to capital, labour and materials will be set off, resulting in an increased demand for labour, capital and materials. As Figure 5-3 demonstrates, all countries, except France in PP20, gain in employment field. Finally, EU-wide employment increases by 112.000 persons in PP10 and 233.000 persons in PP20 for year 2010.

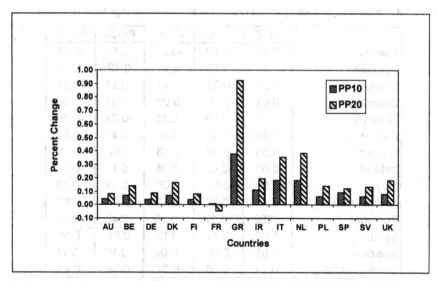

Figure 5-3: Employment changes in year 2010.

One interesting fact that derives from studding the "permit closure" scenarios is that the real wage rate does not increase, as it will happen in the case of double dividend, which is presented below. A reasonable explanation for this would be that the increase in labour demand in the double dividend case is much higher than the increase in the permits case. In the latter case, the decline of labour supply caused by the increase of leisure exceeds the increase of labour demand, which takes place because of the substitution away from energy. On the other hand, substitution processes between inputs and losses in production are responsible for a drop in energy consumption by an average, for all countries, of 5.5% for PP10 and 13.0% for PP20 in year 2010.

Table 5-4: Investment by country for all cases in year 2010 (% change from baseline).

	PP-5	PP-10	PP-15	PP-20	PP-25
Austria	0.00	-0.02	-0.06	-0.17	-0.28
Belgium	0.09	0.15	0.19	0.12	0.11
Germany	0.01	0.02	-0.01	-0.14	-0.26
Denmark	0.04	0.06	0.06	-0.02	-0.06
Finland	-0.02	-0.06	-0.12	-0.24	-0.36
France	-0.05	-0.12	-0.22	-0.41	-0.62
Greece	0.05	0.09	0.13	0.07	0.11
Ireland	0.00	-0.02	-0.05	-0.17	-0.26
Italy	0.01	0.01	0.00	-0.08	-0.13
Netherlands	0.13	0.27	0.42	0.47	0.63
Portugal	-0.09	-0.19	-0.30	-0.46	-0.61
Spain	0.06	0.11	0.12	0.04	0.01
Sweden	0.01	0.00	-0.02	-0.10	-0.17
Un. Kingdom	0.14	0.26	0.37	0.34	0.41

Besides the substitution effect away from energy to labour, there is also a substitution effect away from energy to capital. Both the sectoral and total results show that investment increases when a permit system is introduced in the economy. Since energy becomes more expensive producers prefer to substitute other production factors for energy. The per country results, as Table 5-4 indicates, are not of the same sign for all countries. Austria, Germany, France and Portugal, among others, present a negative investment in all cases. On the contrary, countries such as Belgium, Greece, Netherlands, Spain and the United Kingdom exhibit an increase in their total investment implying a higher elasticity between energy and capital.

Table 5-5: Purchase (+) and sales (-) of permits for PP10 in year 2010.

	AU	BE	GE	DK	FI	FR	GR	IR	IT	NL	PO	SP	SV	UK	EU
Agriculture	0	-1	0	-1	0	1	-1	0	0	4	0	0	-2	-1	0
Coal	0	-3	-4	0	0	-5	0	0	-5	5	0	-2	0	-3	-17
Oil	1	4	11	1	0	10	0	0	8	34	0	5	1	18	93
Gas	1	0	3	1	-3	0	4	1	0	1	0	0	0	4	12
Electricity	-9	-19	-94	-20	-10	-64	-15	-2	-34	-14	-1	-50	-18	-116	-464
Ferrous	-8	-60	-72	0	-6	-91	-1	0	-90	-22	0	-35	-1	-41	-427
Chemical	-11	2	-8	0	0	9	-1	-9	6	17	-1	-4	-6	3	-5
Energy intensive ind.	0	-5	-13	-3	0	-19	-2	-4	-19	2	-2	-13	-8	-23	-109
Electrical goods	0	0	0	0	0	0	0	0	0	0	0	0	0	-2	-2
Transport equipment	0	0	-1	0	0	-1	0	0	0	0	0	0	0	-3	-5
Equipment goods ind.	0	0	1	0	0	0	0	0	0	0	0	0	0	-1	-1
Consumer goods ind.	0	0	1	-2	0	-2	0	-2	3	2	0	-1	0	-7	-8
Building and constract.	0	0	-1	0	0	1	0	0	1	0	0	0	0	-14	-13
Telecomunications	0	0	0	0	0	0	0	0	0	0	0	0	0	0	0
Transport	0	3	11	3	1	14	1	0	21	8	1	10	3	15	92
Services	0	0	0	0	0	0	0	0	0	0	0	0	0	-2	-2
Other market services	0	-2	-2	0	0	-2	0	0	5	3	0	-1	0	-6	-6
Non market services	-1	-1	-4	0	0	-3	0	-1	1	0	0	0	0	-14	-23
Households	21	32	239	18	10	198	5	7	115	31	6	58	36	111	886
Net EU trade volume	-7	-50	68	-3	-8	44	-10	-8	12	71	1	-33	6	-84	0

The sectoral analysis shows that in all countries electricity and the energy intensive industries sell permits. Less emission intensive industries such as transport or households are mostly better off if they decide to buy permits instead of practising substitution. Households, in particular, are those that buy the most permits. As Table 5-5 demonstrates, households buy 886 MECU of permits when a constraint of 10% emission reduction in CO_2 is introduced in the economy. The utility that they obtain by purchasing permits is greater than the utility obtained by substituting other goods for energy. One reasonable explanation for this would be that households are not able to substitute durable goods immediately after the constraint is imposed since demand for energy intensive durable goods is highly inelastic. On the contrary, the energy intensive industries find the substitution less costly than buying permits so they prefer to sell them in the market and reduce their emissions using various ways in order to achieve it. In other words, the cost of reducing emissions below their endowment is less than the benefit from selling the permits to the market.

Summarising, a permit system induces small but important effects in the EU economy. The macro-economic implications, under the permit closure, confirm a positive effect in the economy as both employment and investment benefit from the adjustment of the use of the production factors. Trade gains add to the overall effects. In addition, the total decrease in the CO_2 emission level raises the total economic and environmental welfare. Finally, the macroeconomic effects are also important in distributional terms regarding sectoral activity, as energy intensive and equipment goods industries decline in favour of consumption goods, transportation and services. The macro economic results in total are shown in Table 5-6.

Table 5-6: Macroeconomic aggregates for pollution permits case in 2010 (% change from baseline).

	PP5	PP10	PP15	PP20	PP25
Gross Domestic Product	-0.06%	-0.16%	-0.34%	-0.61%	-0.98%
Employment*	57	112	175	233	283
Private Ivestment	0.03%	0.04%	0.01%	-0.06%	-0.19%
Private Consumption	-0.30%	-0.67%	-1.20%	-1.92%	-2.89%
Domestic Demand	-0.22%	-0.48%	-0.83%	-1.27%	-1.83%
Exports in volume	-0.05%	-0.17%	-0.31%	-0.47%	-0.59%
Imports in volume	-0.88%	-1.82%	-2.95%	-4.26%	-5.79%
Intra trade in the EU	-0.04%	-0.15%	-0.29%	-0.42%	-0.51%
Energy consumption in volume	-2.72%	-5.52%	-8.77%	-12.39%	-16.39%
Consumers' price index	0.33%	0.72%	1.20%	1.81%	2.56%
GDP deflator in factor prices	-0.04%	-0.08%	-0.19%	-0.37%	-0.70%
Real wage rate	-0.32%	-0.70%	-1.19%	-1.85%	-2.70%
Tax revenues as % of GDP***	0.24%	0.52%	0.89%	1.38%	2.01%
Current account as % of GDP***	0.10%	0.21%	0.34%	0.48%	0.67%
Marginal abatement cost (ECU'85/tn C)	45.6	100.9	176.2	277.3	414.4
Total atmospheric emissions					
CO2	-5.04%	-9.84%	-14.72%	-19.62%	-24.56%
NOX	-3.66%	-7.58%	-11.79%	-16.21%	-20.85%
SO2	-8.12%	-15.17%	-21.87%	-28.07%	-33.82%
VOC	-1.72%	-4.27%	-7.28%	-10.68%	-14.44%
PM	-9.19%	-17.09%	-24.46%	-31.20%	-37.36%

* in thousand employed persons
** as percent of GDP (last simulation year)
*** absolute difference from baseline

6 The Role of Accompanying Policies

The emission-constrained cases impose additional costs to all producers and consumers. This additional cost, being necessary to reach the target, depends on the equilibrium shadow cost per unit of emissions, which is computed at the EU level. Mathematically, the mechanism is exactly equivalent to the imposition of that level of carbon tax (uniform across the EU) that is necessary to reach the target.

6.1 Recycling through the Reduction of Labour Costs

The full version of *GEM-E3* was utilised to run a double dividend application for the EU-14 member-states regarding the environment (CO_2) and employment. This version considers full competitive equilibrium in all markets, including goods, labour and capital markets.

6.1.1 Scenario Assumptions

The assumptions for the Double Dividend (*DD*) case are partly the same with those used for the Pollution Permit case. However, it is further assumed in this scenario that public finances collect the counterpart of all the additional costs implied by the carbon constraint and then recycle them by reducing labour cost (social security rate paid by the employers). This corresponds to the so-called double dividend analysis, expecting that benefits will arise both for the environment and employment.

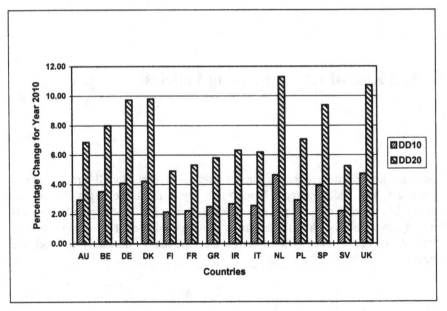

Figure 6-1: Reduction in social security rate.

Regarding the accompanying reduction of the rate of social security contribution of employers, the following should be noted:

- this reduction operates uniformly in all sectors, independently of their relative labour costs or the pre-existing level of the rate;

- the level of reduction in the rate of social security contribution of employers that is necessary to compensate exactly for the additional revenues that are determined ex-post according the shadow price of the environmental constraint. The model iterates among the possible values of rate reduction and determines the one needed in the new equilibrium achieved. Figure 6-1 shows the ex-post values obtained for DD10 and DD20.

6.1.2 Is there a Positive Double Dividend?

Figure 6-2, Figure 6-3, and Figure 6-4 show that a positive double dividend exists for all EU-14 member-states. As expected, the benefits for the environment (measured by the reduction of damages) are by far larger than the benefits for employment. In the EU-14, total employment increases by around 638.000 persons, in DD10 while total CO_2 emissions decrease by 10 % in the steady-state for the EU as a whole. In addition, NO_x and SO_2 emissions also decrease, namely by 7.5 % and 15.62 % respectively. This is not due to abatement technology, but can be explained by the improvement of energy efficiency and the changes in the fuel mix. In the DD20 an even larger increase in employment is observed, 1.460.000 persons, showing that the increase in employment does not follow a linear trend with respect to the decrease in emissions of CO_2. This also can be observed by the DD5, DD15 and DD25. Each country experiences different employment gains. Spain, as Figure 6-3 shows, is the country that gains the most in employment. This implies that the labour elasticity in this country is higher. On the contrary, the United Kingdom gains less in employment from in these scenarios. This can be explained from two facts. On one hand, the UK is a country that faces smaller unemployment, which explains the smaller increase in employment gains. On the other hand, the United Kingdom is an oil producer and thus is less affected by the decrease in energy demand.

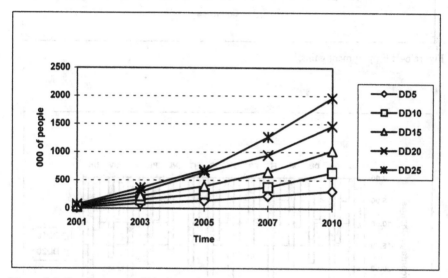

Figure 6-2: Employment in 000 of people.

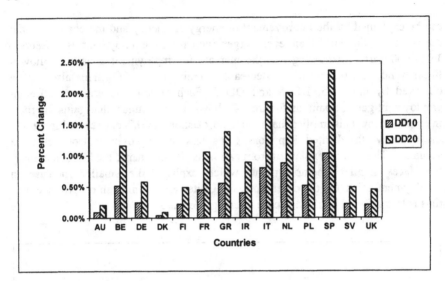

Figure 6-3: Employment gains.

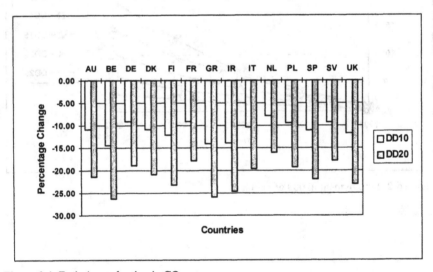

Figure 6-4: Emission reduction in CO_2.

Table 6-1: Energy consumption for the year 2010.

	DD-5	DD-10	DD-15	DD-20	DD-25
Austria	-3.02	-6.23	-9.85	-13.80	-18.06
Belgium	-3.55	-7.16	-11.11	-15.40	-20.07
Germany	-3.37	-6.98	-11.06	-15.53	-20.37
Denmark	-3.78	-7.72	-12.12	-16.90	-22.04
Finland	-3.28	-6.65	-10.32	-14.23	-18.35
France	-2.70	-5.55	-8.76	-12.32	-16.26
Greece	-3.88	-7.88	-12.24	-16.89	-21.81
Ireland	-3.07	-6.21	-9.68	-13.46	-17.52
Italy	-3.06	-6.30	-9.97	-14.06	-18.60
Netherlands	-2.12	-4.51	-7.52	-11.30	-16.01
Portugal	-2.78	-5.84	-9.36	-13.31	-17.65
Spain	-3.01	-6.28	-10.00	-14.07	-18.49
Sweden	-2.78	-5.80	-9.28	-13.20	-17.53
Un. Kingdom	-3.98	-8.05	-12.51	-17.28	-22.31

The application of the double dividend policy, as defined above, initiates an increase in energy prices coupled with a reduction in labour costs. Prices of production factors change non uniformly, resulting in substitutions in favour of labour and away from energy. Energy demand decreases more in the sectors that are heavy consumers of energy (up to -6,3% in that sector for the DD10 and up to 14,2% for the DD20, in year 2010). In general, electricity and gas substitute the other energy forms, as expected. The fuel switching is however moderate in all sectors. Energy consumption decreases, differ by country, ranging from 4.5% to 8% in DD10 and for 12% to 17% in DD20. The imposed decrease of CO_2 emissions is of the order of 10% and 20% respectively, while substantial gains on the other pollutants estimated by the model (SO_2 and NOx) are also obtained. Table 6-1 shows total energy consumption for year 2010.

By construction, the surplus or deficit of public budget remains unchanged. Thus, this case does not alter, in a first approximation, income distribution and does not imply any need for public budget financing (or any new financing capacity from the public budget). Income distribution is affected indirectly, through the increase in the value of labour compared to other factors and the implied mechanism of income distribution, through the Social Accounting Matrix of the model.

6.1.3   Indirect and Equilibrium Effects

As mentioned, the assumed double dividend policy entails a change of relative costs of production factors. The unit cost of labour decreases, while the cost of energy increases. Labour becomes more competitive, not only compared to energy, but also with respect to other production factors including capital. Thus, labour substitutes for energy but also for capital. This implies a diminishing rate of return of capital and a slowdown of investment.

To this effect (technology optimisation), one must add a demand effect. The production costs bear different charges by sector, due to different energy and labour intensities. These varying costs, after being reflected into prices, will imply substitutions both in final and intermediate demand. Those sectors that face a diminishing market have an additional reason to invest less: due to the limited substitutability of the factors and the fixed capital constraint the rate of return on capital tends to further decrease.

The sectors, on the other hand, that see their market augmenting will tend to invest more, since they will attribute a higher shadow price to their capital stock, as they desire more production capacity. In this case, the demand effect counteracts the substitution effect. Thus, the overall outcome regarding investment is uncertain. In the GEM-E3 model however it tends to be negative for almost all sectors. Table 6-2 shows this decrease in investments in all countries for all cases.

Table 6-2: Investment by country (% change from baseline).

	DD-5	DD-10	DD-15	DD-20	DD-25
Austria	-0.12	-0.26	-0.46	-0.71	-1.02
Belgium	-0.10	-0.24	-0.44	-0.71	-1.08
Germany	-0.07	-0.17	-0.31	-0.51	-0.78
Denmark	-0.13	-0.29	-0.50	-0.78	-1.12
Finland	-0.08	-0.19	-0.33	-0.51	-0.73
France	-0.10	-0.22	-0.38	-0.60	-0.87
Greece	-0.25	-0.53	-0.88	-1.31	-1.84
Ireland	-0.22	-0.48	-0.79	-1.19	-1.68
Italy	-0.09	-0.20	-0.34	-0.53	-0.77
Netherlands	0.01	0.00	-0.04	-0.14	-0.34
Portugal	-0.20	-0.42	-0.68	-0.99	-1.35
Spain	-0.11	-0.25	-0.45	-0.70	-1.02
Sweden	-0.11	-0.24	-0.42	-0.64	-0.92
Un. Kingdom	-0.06	-0.15	-0.29	-0.51	-0.80

The increase in real wages pushes private income and consumption, which keeps up domestic demand. Because of higher production costs, the competitiveness of domestic firms tend to deteriorate and the volume of exports shrinks (Table 6-3). The volume of non energy imports from the Rest of the World (RoW) increases.

Shifts of total domestic demand are then uncertain, since they are positively influenced by private consumption and negatively influenced by trade. Shifts of total investment are also uncertain, for the reasons explained above. Of course, the energy intensive industries suffer the most, facing the highest decrease of their market. Non energy intensive sectors face, on the other hand, an increasing demand.

Total imports remain unchanged or decrease, because the decrease in mostly imported energy-goods compensates the increase of non energy imports, which is due to competitiveness losses. As it can be observed from the sectoral results of Table 6-4 the sectors that experience a decrease in imports are the equipment good industries (2.1% for DD10 and 3.6% for DD20 for year 2010) and all the energy sectors. As mentioned, the current account at the country level can adjust (is not fixed). The EU member-states are interdependent through flexible bilateral trade flows. The behaviour of the Rest of the World (RoW) is assumed exogenous. Then the EU countries, by definition, do not face a price reaction of the RoW when they demand more imports. The EU countries do so because of the higher production costs that are due to imperfect adjustment of production technology to the changes of relative production factor prices. The absence of readjustment of RoW imports combined with the unchanging RoW export prices, entail a revaluation of EU exports that will increase more than the value of imports. The latter can even decrease, depending on the elasticities and the flexibility of shifting away from energy, which is basically imported in the EU. This imply an improvement of the terms of trade[26] for the EU, or in other terms a willingness of the RoW to accept a deterioration of their terms of trade. It turns out that this willingness is a necessary condition for a double dividend.

---

[26] Terms of trade are defined here (as it is standard in the literature), as the ratio of export price over import price.

Table 6-3: Volume of exports by country (% change from baseline).

	DD-5	DD-10	DD-15	DD-20	DD-25
Austria	-1.27	-2.62	-4.15	-5.88	-7.82
Belgium	-0.91	-1.91	-3.08	-4.42	-5.95
Germany	-1.24	-2.62	-4.26	-6.16	-8.35
Denmark	-1.52	-3.11	-4.90	-6.91	-9.18
Finland	-0.89	-1.82	-2.88	-4.07	-5.42
France	-0.45	-0.94	-1.52	-2.19	-2.95
Greece	-0.69	-1.57	-2.69	-4.05	-5.67
Ireland	-0.58	-1.18	-1.86	-2.63	-3.50
Italy	-0.30	-0.64	-1.07	-1.59	-2.22
Netherlands	-1.01	-2.16	-3.61	-5.42	-7.66
Portugal	-0.25	-0.58	-1.00	-1.53	-2.21
Spain	-0.69	-1.54	-2.63	-3.97	-5.60
Sweden	-0.71	-1.46	-2.35	-3.36	-4.52
Un. Kingdom	-1.81	-3.64	-5.66	-7.89	-10.32

Since prices increase all over the European Union, it is understandable that intra-EU volume of trade decreases much more than the marginal reduction of the exports of the rest-of-the-world to the EU. The result of this, is a small increase in the import dependency of the EU. The decrease of intra-EU trade is, once again, not uniform across countries and sectors. The different export prices set by each sector in every country lead some to become more or less competitive and thereby increase or decrease respectively their export shares. The country that benefits the most from the adjustments of trade is the United Kingdom, which minimises its export losses.

Table 6-4: Volume of imports by sector (% change from baseline).

	DD-5	DD-10	DD-15	DD-20	DD-25
Agriculture	0.41	0.84	1.31	1.79	2.32
Coal	-11.02	-20.60	-29.21	-36.65	-43.01
Crude oil and oil products	-1.87	-4.00	-6.56	-9.51	-12.84
Natural gas	-1.48	-3.27	-5.63	-8.70	-12.56
Electricity	-1.15	-2.46	-4.02	-5.83	-7.89
Ferrous, non-ferrous	1.48	3.12	5.09	7.43	10.16
Chemical products	1.39	2.98	4.94	7.37	10.41
Other energy intensive ind.	0.98	2.06	3.33	4.81	6.51
Electrical goods	0.37	0.71	0.99	1.37	1.55
Transport equipment	0.59	1.19	1.84	2.50	3.14
Other equipment goods ind.	-1.23	-2.10	-2.71	-3.06	-3.23
Consumer goods industries	0.19	0.36	0.52	0.68	0.87
Building and construction	-0.25	-0.62	-1.34	-2.72	-4.97
Telecommunication services	-0.13	-0.32	-0.61	-1.04	-1.63
Transports	0.78	1.71	2.89	4.36	6.13
Credit and insurance	1.29	2.73	4.50	6.68	9.38
Other market services	0.75	1.51	2.30	3.09	3.86
Non market services	0.37	0.71	0.99	1.32	1.43

The combined effects from changes in supply, domestic demand and unit costs of labour may compensate each other. In the present simulation, small increases of domestic prices and inflation are observed.

Regarding effects on total activity, the positive effects of demand are found to over-compensate negative effects from trade and thus GDP in factor prices is slightly higher.

6.1.4   Summary of Conditions for a Double Dividend

The results and sensitivity analyses confirm the following points:

- the positive double dividend result, is achieved everywhere. There are increases in employment and in real wages, while energy consumption falls significantly leading to fewer emissions for CO_2 and other pollutants;

- the magnitude of the employment dividend depends crucially on the labour market regime. An inelastic labour supply reduces the gain in employment and results in higher competitiveness losses and inflationary pressures;

- distributional effects across sectors are important. The energy intensive industry suffers the most, followed by the equipment goods industry. On the other hand, consumer goods and services are favoured from the redirection of demand towards them;

- the decrease in competitiveness of the EU-14, reflected in the decrease of both intra-EU trade and in the volume of exports to the rest-of-the-world, is overcompensated by the significant improvement of the EU terms of trade, leading to an increase in the money inflow from exports;

- a double dividend is only obtained if the rest-of-the-world accepts the deterioration of its terms of trade (in other words, it is willing to buy more expensive products) with the EU member states, so as to support the amelioration of the current account of the EU.

The effects differ significantly among countries. This is attributed to the different structure of EU economies, especially regarding:

- the flexibility of the labour market (demand and supply);

- the degree of exposure of sectors to foreign trade and the dependence of the economy on sectors that are affected by the policy ;

- the pre-existing level of energy-related excise taxes;

- the flexibility of the energy supply system (mainly power generation) to adapt and the possibilities for fuel switching (in particular to natural gas).

Sensitivity analysis confirmed that the differences among countries due to the structure of the economy than to the values of any particular elasticities. In any case, the distributional effects among countries are not neutral and they must be taken into account if the policy is to be implemented at the European Union level. Table 6-5 shows the macro-economic data obtained by implementing the "Double Dividend" case.

Table 6-5: Macroeconomic aggregates for labour recycling in 2010 (% changes from baseline).

	DD5	DD10	DD15	DD20	DD25
Gross Domestic Product	-0.08%	-0.20%	-0.38%	-0.65%	-1.01%
Employment*	305	638	1023	1460	1961
Private Ivestment	-0.08%	-0.19%	-0.35%	-0.56%	-0.84%
Private Consumption	0.30%	0.57%	0.80%	0.99%	1.10%
Domestic Demand	-0.26%	-0.57%	-0.94%	-1.40%	-1.97%
Exports in volume	-0.98%	-2.09%	-3.39%	-4.90%	-6.63%
Imports in volume	-0.48%	-0.99%	-1.57%	-2.18%	-2.82%
Intra trade in the EU	-1.01%	-2.13%	-3.46%	-4.99%	-6.74%
Energy consumption in volume	-3.18%	-6.55%	-10.35%	-14.55%	-19.15%
Consumers' price index	1.00%	2.13%	3.51%	5.18%	7.23%
GDP deflator in factor prices	0.05%	0.10%	0.15%	0.18%	0.17%
Real wage rate	0.23%	0.43%	0.61%	0.78%	0.92%
Tax revenues as % of GDP***	0.79%	1.66%	2.70%	3.94%	5.41%
Current account as % of GDP***	-0.04%	-0.09%	-0.15%	-0.22%	-0.30%
Marginal abatement cost (ECU'85/tn C)	39.1	88	153.3	240	356.6
Total atmospheric emissions					
CO2	-5.00%	-10.00%	-15.00%	-19.97%	-24.95%
NOX	-3.56%	-7.55%	-11.74%	-16.10%	-20.62%
SO2	-8.12%	-15.62%	-22.71%	-29.28%	-35.38%
VOC	-1.52%	-3.94%	-6.73%	-9.84%	-13.25%
PM	-9.03%	-17.29%	-24.97%	-31.98%	-38.42%

* in thousand employed persons
** as percent of GDP (last simulation year)
*** absolute difference from baseline

6.2 Recycling through the Reduction of Investment Costs

One of the main negative implications of the policy scenario presented above, is the reduction of investment. This results in a decrease in the productive capacity of the sectors of the EU countries. This is an additional factor that leads to upward pressures in prices. In the longer term, the dynamic effects on production, competitiveness and growth, not captured in the simulation presented above, may be negative, leading to a decline in employment gains.

An alternative approach would be to use the revenues from the CO_2 tax, to subsidise investments. This way, it can be expected that the expansion of the capital stock will relax the capacity constraints in the economy, thereby leading to lower prices and increased competitiveness. More explicitly, the amount collected by public finance, corresponding to the additional costs, is used to subsidise the cost of investment of firms. This can be interpreted as a relaxation in the financing constraints of the economy, leading to lower lending rates for private investment.

98

6.2.1   Scenario Assumptions

Within the "Capital Recycling" closure of the model, it is assumed that the authorities collect the counterpart of all additional costs, resulted from the environmental constraint and recycle them through reducing the cost of investment

The comparison of the stock of capital in the current year with the desired stock of capital in the next year determines the investment decision of the firms. This decision is also affected by the price of capital and the cost of investment required obtaining it. In the present scenario the public sector reduces the price of the investment by distributing the revenues collected from the imposition of the emission reduction constraint. As was mentioned above the additional cost of the emission reduction constraint takes the form of an endogenous tax. In order to provide the magnitude of the importance of the additional costs, Figure 6-5 shows total revenues as a % of GDP, from the endogenous CO_2-tax as a result of the reduction in emissions. These range from about 1.0 to 2.0% of GDP for CR10 and 2.3 to 4.8% for CR20, being very similar to the range derived from the "double dividend" case.

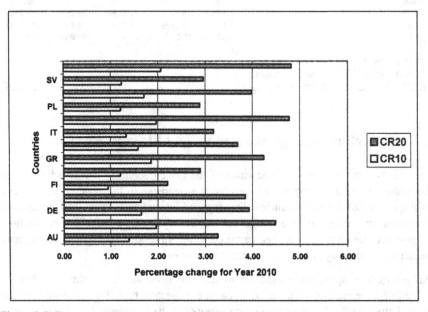

Figure 6-5: Tax revenues as percentage of GDP.

6.2.2 Equilibrium Effects

The main findings of the capital recycling approach in EU-14 are shown in Figure 6-6 and Figure 6-7. Total employment decreases slightly by around 38.000 persons in CR10. In addition, NO_x and SO_2 emissions also decrease, namely by 7.62 % and 15.35 % respectively. This is not due to abatement technology, but to the improvement of energy efficiency and the changes in fuel mix. In the CR20 an even larger decrease in employment is observed, amounting to 92.000 persons, showing that the decrease in employment does not follow a linear trend with respect to the decrease in emissions of CO_2. This can also be observed in Figure 6-6 where all the scenarios are presented. Compared with the double dividend case, where the gains in employment are positive for all countries, in this scenario the results are mixed. Some countries, such as Sweden, Denmark, Finland and the United Kingdom experience gains while others experience a decline in employment. Given the nesting in of the production structure, the fall of the cost of capital makes it more attractive than labour, energy and materials. Thus, labour tends to be substituted for capital, although at a lesser degree than energy. It is obvious that countries such as Spain and Belgium have higher elasticities of substitution between labour and capital, which lead to high employment losses.

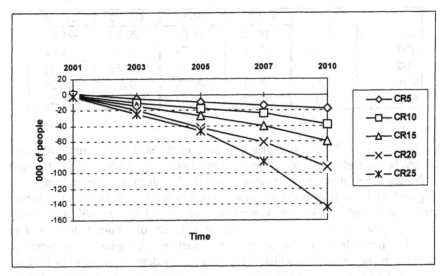

Figure 6-6: Employment in 000 of people.

100

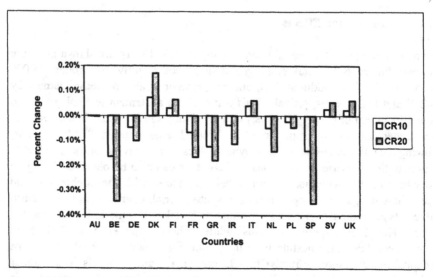

Figure 6-7: Employment changes in 2010.

Table 6-6: Total emissions for year 2010 (% change from baseline).

	CR-5	CR-10	CR-15	CR-20	CR-25
CO2	-5.00	-10.00	-15.00	-20.00	-25.00
NOX	-3.56	-7.55	-11.74	-16.10	-20.62
SO2	-8.12	-15.62	-22.71	-29.28	-35.38
VOC	-1.52	-3.94	-6.73	-9.84	-13.25
PM	-9.03	-17.29	-24.97	-31.98	-38.42

Companies, through their optimising behaviour, demand more capital and less energy, labour and materials. Within a competitive labour market regime, the shift of the aggregate labour demand curve would imply a decline of the real wage rate (Figure 6-8). Indeed, the real wage rate is found to decrease in all countries except the United Kingdom. This can be explained by the fact that the UK is a country that experiences low unemployment so the elasticity of labour with respect to capital is rather low. On the contrary, the elasticity of capital with respect to energy is higher in the UK, which reduces its energy demand the most (8.14% for CR10 and 17.48% in CR20) in relation to other countries of the European Union. Sweden, experiences the highest decrease in the wage rate (-2.7% for CR20 relative to 0.4% of the United Kingdom) and the smallest decrease in energy demand (-12.3% for CR20). Figure 6-8 shows that the countries that are heavy consumers of energy such as Germany, Netherlands and the United Kingdom are those that experience the smallest decrease in the real wage rate.

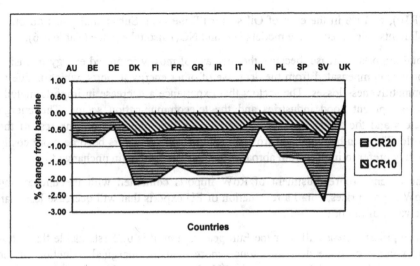

Figure 6-8: Real wage rate in (% change from baseline).

Table 6-7: Energy demand by sector for EU (year 2010).

	CR-5	CR-10	CR-15	CR-20	CR-25
Agriculture	-3.42	-7.17	-11.43	-16.15	-21.36
Coal	-11.66	-21.96	-31.43	-39.90	-47.58
Crude oil and oil products	-2.05	-4.41	-7.21	-10.47	-14.27
Natural gas	-3.09	-6.55	-10.63	-15.36	-20.90
Electricity	-4.36	-8.68	-13.12	-17.59	-22.16
Ferrous, non-ferrous	-6.22	-12.29	-18.45	-24.53	-30.64
Chemical products	-2.46	-5.24	-8.55	-12.40	-16.94
Other energy intensive ind.	-2.66	-5.54	-8.80	-12.40	-16.47
Electrical goods	-1.98	-4.20	-6.75	-9.50	-12.67
Transport equipment	-2.19	-4.66	-7.52	-10.62	-14.23
Other equipment goods ind.	-1.90	-3.99	-6.38	-8.99	-11.90
Consumer goods industries	-2.29	-4.84	-7.78	-11.11	-14.87
Building and construction	-3.32	-6.95	-11.03	-15.46	-20.29
Telecommunication services	-2.66	-5.60	-8.95	-12.68	-16.84
Transports	-2.88	-6.16	-10.02	-14.43	-19.46
Credit and insurance	-3.24	-6.72	-10.57	-14.73	-19.24
Other market services	-2.69	-5.66	-9.06	-12.84	-17.08
Non market services	-2.91	-6.10	-9.68	-13.61	-17.92

The application of the capital recycling policy, as defined above, initiates an increase in energy prices coupled with a reduction in capital costs. Prices of production factors change non uniformly, resulting in substitutions in favour of capital and away from energy, labour and materials. Energy demand decreases more in the energy intensive sectors. The energy sectors such as Coal, Oil, Gas and Electricity experience a reduction ranging from 21.9% of the case of Coal

102

(CR10), to 14% in the case of Oil sector (Table 6-7). Substantial gains on other pollutants calculated by the model (SO_2 and NO_x) also take place (Table 6-6).

Total imports decrease, because the increase of mostly imported energy intensive goods is compensated from the decrease of other energy imports, which is due to competitiveness losses. The sectors that experience a decrease in imports include the equipment good industries and the telecommunication sector. The energy sectors and the transports face an increase in their imports from the rest of the world. This can be explained from the fact that these sectors become more costly so they prefer to import from abroad where the prices remain unchanged.

The absence of readjustment of RoW imports combined with the unchanging RoW export prices, entail a revaluation of EU exports that will decrease less than the value of imports.

Since prices increase all over the European Union, it is understandable that intra-EU volume of trade decreases much more than the marginal reduction of the exports of the rest-of-the-world to the EU. The country that benefits the most from the adjustments of trade is the United Kingdom, which is the only country that actually increases its exports.

The combined effects from supply, domestic demand and unit costs of capital may compensate each other. As a matter of fact, small increases of domestic prices and inflation are observed. All countries except Germany for CR10 and the United Kingdom for both cases face an increase in their consumer price indices. Demand push is moderated by the flexibility of labour supply, while demand contraction in some sectors moderates price increases due to cost increases (Figure 6-9).

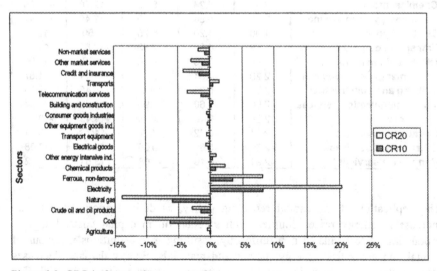

Figure 6-9: GDP in factor prices.

The overall impact of this scenario on activity is ambiguous and depends on the specific characteristics of each country. The energy intensive sectors such as the electricity, ferrous and non ferrous, the chemicals and other energy-related industries face an increase in their factor prices while others face a decline in their prices.

6.2.3 Subsidising Investments

The most important aspect of this scenario run is the effects that the investment subsidy will have on the capital accumulation in particular and the economic activity as this forms dynamically, in general. In the double dividend case the government recycles the revenues collected from the tax on CO_2 by reducing the cost of labour. It is natural that the production factor that benefits the most will be labour. In the present simulation, it is the capital that benefits which results in an investment increase.

In the short run, domestic economic activity is depressed, as households' disposable income decreases in real terms, because of the fall in real wages. Investment increases are not enough to compensate for the fall in private consumption. The capital recycling policy leads to an increasing rate of return of capital and an increase in investment. However, investment is also affected by changes in demand. The sectors, for which demand has declined, like energy intensive industries, will obviously reduce investment spending. Other sectors will experience an increase in demand and will expand their investment for reasons other than the cost of capital. Figure 6-10 and Table 6-8 show the changes by country and by sector respectively.

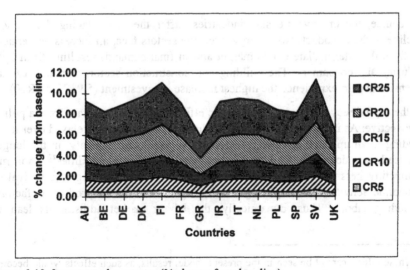

Figure 6-10: Investment by country (% change from baseline).

Households consider simultaneously the propensity to invest more in durable goods because of lower costs, but also the operating costs of durable goods that become higher because of higher energy prices. This combination turns out to be positive (in this model application) for the demand of durable goods. This effect must be added to the effect from the increase of total investment (as explained), entailing an even higher increase of the demand for equipment goods.

Table 6-8: Investment by sector.

	CR-5	CR-10	CR-15	CR-20	CR-25
Agriculture	0.52	1.05	1.68	2.39	3.20
Coal	-8.86	-16.85	-24.32	-31.07	-37.21
Crude oil and oil products	-0.66	-1.46	-2.43	-3.52	-4.73
Natural gas	-1.10	-2.44	-4.14	-6.20	-8.71
Electricity	0.44	0.93	1.52	2.24	3.09
Ferrous, non-ferrous	0.17	0.35	0.59	0.94	1.38
Chemical products	0.65	1.33	2.11	3.00	4.00
Other energy intensive ind.	0.87	1.81	2.91	4.23	5.75
Electrical goods	0.98	2.03	3.26	4.79	6.53
Transport equipment	0.96	1.96	3.12	4.55	6.10
Other equipment goods ind.	1.05	2.20	3.58	5.26	7.26
Consumer goods industries	0.75	1.56	2.50	3.61	4.88
Building and construction	1.21	2.53	4.08	5.93	8.10
Telecommunication services	0.42	0.84	1.32	1.85	2.42
Transports	0.68	1.40	2.25	3.23	4.37
Credit and insurance	0.40	0.80	1.26	1.78	2.35
Other market services	0.43	0.86	1.35	1.90	2.49
Non market services	0.49	1.01	1.62	2.33	3.15

Of course, the energy intensive industries suffer the most, facing the highest decline in their product. Non energy intensive sectors face, an increasing demand (both in the intermediate goods market and in final demand) resulting from both industry and consumers. The building and construction sector, a mainly capital intensive sector, experience the highest increase in investment (5.9% for CR20).

In the capital recycling case, the dynamic effects are most important from a policy perspective: As investment increases, production constraints are relaxed over time, leading to an upwards shift of the economy's production capacity, in the longer run leading to decreasing equilibrium prices of domestic commodities[27]. Short run competitiveness losses are compensated after some time, leading to a revival of economic activity. This implies gradually higher demand for production factors, including labour. Thus, dynamically, the labour market adjustments lead to

---

[27] These effects cannot be seen in the present model results, as such effects would become apparent after 2010.

cancelling out losses in real wages, that further supports disposable income and private consumption. The combined effects of these movements will imply higher GDP and more employment in the long run.

Finally, Table 6-9 presents the macroeconomic aggregates in year 2010 for all scenarios.

Table 6-9: Macroeconomic aggregates with investment recycling for EU-14 in year 2010 (% change from baseline).

	5%	10%	15%	20%	25%
Gross Domestic Product	-0.08%	-0.21%	-0.40%	-0.66%	-1.03%
Employment*	-18	-38	-60	-92	-144
Private Ivestment	0.51%	1.04%	1.65%	2.37%	3.18%
Private Consumption	-0.44%	-1.00%	-1.70%	-2.58%	-3.75%
Domestic Demand	-0.37%	-0.80%	-1.31%	-1.91%	-2.66%
Exports in volume	-0.26%	-0.58%	-0.97%	-1.40%	-1.87%
Imports in volume	-1.02%	-2.12%	-3.37%	-4.75%	-6.32%
Intra trade in the EU	-0.24%	-0.53%	-0.87%	-1.26%	-1.67%
Energy consumption in volume	-3.08%	-6.37%	-10.05%	-14.07%	-18.53%
Consumers' price index	0.09%	0.22%	0.42%	0.69%	1.07%
GDP deflator in factor prices	-0.74%	-1.56%	-2.51%	-3.61%	-4.94%
Real wage rate	-0.27%	-0.62%	-1.06%	-1.61%	-2.36%
Tax revenues as % of GDP***	0.74%	1.58%	2.56%	3.72%	5.12%
Current account as % of GDP***	0.11%	0.22%	0.35%	0.49%	0.67%
Marginal abatement cost (ECU'85/tn C)	36.6	81.8	140.7	216.5	316.9
Total atmospheric emissions					
CO2	-4.92%	-9.95%	-14.95%	-19.92%	-24.96%
NOX	-3.57%	-7.62%	-11.84%	-16.20%	-20.78%
SO2	-7.92%	-15.35%	-22.38%	-28.89%	-35.03%
VOC	-1.69%	-4.29%	-7.23%	-10.45%	-14.01%
PM	-8.82%	-17.05%	-24.71%	-31.71%	-38.24%

* in thousand employed persons
** as percent of GDP (last simulation year)
*** absolute difference from baseline

6.3 No Recycling

In this scenario (NR), it is assumed that all additional costs from the imposed constraint are collected by the authorities and are not further distributed. All agents face increasing costs of energy use without any compensation. In this case, the emission reduction target acts very restrictively to the economy, leading to recession. This can be explained by the fact that the amount corresponding to the additional costs could be otherwise consumed or invested.

In case of non recycling, the -20% target (compared to baseline) in 2010 has significant macro-economic implications. The Gross Domestic Product of the EU is found reduced by -1.1% in 2010, consumption by -2.85% unemployment increased by 531 thousand in 2010. Consumer prices, decrease by -0.37% in 2010, leading to some preservation of exports to the rest-of-the-world similar to their level in the baseline) and a fall of imports. Because of the fall of labour demand, the real wage rate decreases as well. The emissions target requires additional costs that amount to 3.64% of the EU GDP in 2010. The shadow price of the constraint is found equal to 208 ECU'85 per ton of carbon. The economic welfare is consequently reduced (-0.58%), while the avoidance of environmental damages cannot compensate this fall, leading to a decrease of total welfare by -0.45%. This policy, under the assumption of non recycling of additional costs, corresponds to a loss of 400 ECU per capita for the whole EU, during the 2000-2010 decade.

The distribution of results among the member-states, depends on the industrial structure and the pattern of use of fossil fuels, reveal some differences. To explain them, one should consider the re-adjustment of intra-EU trade (which increases by 0.33% in 2010), serving to partly alleviate the effects. GDP reduction effects range from -0.7% in 2010 (compared to -1.1% for whole EU), observed for France, Sweden, Finland, up to -1.6% and beyond, observed for Belgium, Netherlands, Greece and Spain. Reduction of emissions ranges from -15% to -25% in 2010 (compared to baseline), which is rather narrow. Total welfare changes depend also on concentrations of acid rain pollutants, as they are found significantly reduced as a side effect of the CO_2 target. Some countries, like Belgium and Netherlands bear significant effects from trade. The overall aggregate macroeconomic results are presented in Table 6-10.

Table 6-10: Macroeconomic aggregates with no recycling for EU-14 in year 2010 (% change from baseline).

	NR5	NR10	NR15	NR20	NR25
Gross Domestic Product	-0.18%	-0.40%	-0.70%	-1.10%	-1.60%
Employment*	-109	-225	-362	-531	-743
Private Investm.	-0.27%	-0.56%	-0.93%	-1.38%	-1.94%
Private Consumption	-0.52%	-1.12%	-1.89%	-2.85%	-4.06%
Domestic Demand	-0.55%	-1.14%	-1.85%	-2.69%	-3.69%
Exports in volume	0.04%	0.05%	0.06%	0.08%	0.14%
Imports in volume	-1.33%	-2.72%	-4.30%	-6.08%	-8.06%
Intra trade in the EU	0.09%	0.16%	0.23%	0.33%	0.49%
Energy consumption in volume	-3.18%	-6.49%	-10.18%	-14.24%	-18.66%
Consumers' price index	-0.13%	-0.23%	-0.32%	-0.37%	-0.38%
GDP deflator in factor prices	-1.03%	-2.12%	-3.38%	-4.84%	-6.56%
Real wage rate	-0.31%	-0.66%	-1.11%	-1.68%	-2.38%
Tax revenues as % of GDP***	0.74%	1.54%	2.50%	3.64%	4.99%
Current account as % of GDP***	0.17%	0.35%	0.56%	0.79%	1.07%
Marginal abatement cost (ECU'85/tn C)	36.1	79.6	135.8	208.0	301.5
Total atmospheric emissions					
CO2	-5.00%	-9.98%	-14.99%	-19.99%	-24.98%
NOX	-3.67%	-7.72%	-11.96%	-16.36%	-20.90%
SO2	-7.96%	-15.29%	-22.28%	-28.82%	-34.92%
VOC	-1.80%	-4.44%	-7.41%	-10.69%	-14.24%
PM	-8.84%	-16.95%	-24.56%	-31.58%	-38.07%

* in thousand employed persons
** as percent of GDP (last simulation year)
*** absolute difference from baseline

6.4 The Impact of Exemptions and Derogations

In this simulation, the double dividend scenario (DD20) has been modified in order to incorporate two different cases of sensitivity analysis. The first scenario case (678-20) examines the effectiveness of exempting the energy intensive sectors (ferrous and non ferrous ore and other metals, chemical products and other energy intensive industries) from the impact of the environmental policy. The second one (GISP-20) examines the case of derogating action in the cohesion countries (Greece, Ireland, Spain and Portugal) for the first two years of the policy implementation. In both cases the same assumptions, which have been made for the double dividend scenario, are applied. The model assumptions concerning the labour market are kept the same so as a robust comparison can be conducted. In the following table the macroeconomic results are presented for the three scenarios.

Table 6-11: Macroeconomic aggregates in year 2010 for 20% reduction in CO_2 emissions.

Macroeconomic Aggregates in 2010			
	DD20	GISP20	678-20
Gross Domestic Product	-0.65%	-0.65%	-0.72%
Employment*	1460	1461	2036
Private Ivestment	-0.56%	-0.56%	-0.52%
Private Consumption	0.99%	0.99%	0.47%
Domestic Demand	-1.40%	-1.41%	-1.57%
Exports in volume	-4.90%	-4.91%	-4.87%
Imports in volume	-2.18%	-2.18%	-4.20%
Intra trade in the EU	-4.99%	-4.99%	-5.05%
Energy consumption in volume	-14.55%	-14.56%	-17.33%
Consumers' price index	5.18%	5.19%	8.09%
GDP deflator in factor prices	0.18%	0.18%	0.19%
Real wage rate	0.78%	0.78%	0.12%
Tax revenues as % of GDP***	3.94%	3.94%	5.51%
Current account as % of GDP***	-0.217%	-0.217%	-0.014%
Marginal abatement cost (ECU'85/tn C)	240.0	240.3	438.0
Total atmospheric emissions			
CO2	-19.97%	-19.98%	-19.98%
NOX	-16.10%	-16.11%	-20.31%
SO2	-29.28%	-29.29%	-28.88%
VOC	-9.84%	-9.85%	-15.18%
PM	-31.98%	-32.00%	-36.33%

* in thousand employed persons
** as percent of GDP (last simulation year)
*** absolute difference from baseline

From the results presented in the above table, it is obvious that the scenario concerning the derogation of South for the first two years of the policy implementation have almost no difference as far as the impact in the economy is concerned. The changes are negligible and no more than 0.02%. It should be mentioned that derogation is an issue that can only partly be covered by GEM-E3. Due to the fact that expectations in GEM-E3 are backward looking the fact that some countries face the tax, will not alter the policy of companies operating in countries that do not have the cost. The only effect that remains, is through trade, where lower prices in countries with no environmental constraint will augment their market, but only for the period of the derogation.

When these countries finally start imposing a constraint on emissions they have to undergo a transitory period, which the other countries have already overcome. So, this policy is likely to lead to some short-term gains followed by longer-term losses.

On the contrary the exception of the heavy industry from the policy has substantial changes compared to the original double dividend scenario. In year 2010 and for 20% reduction in CO_2, the decline in GDP is 0.07% more than the one achieved in the other two scenarios. However it should be mentioned by observing Figure 6-11 below, that there is a turning point in year 2007. In the years before 2007 the decline in the GDP for the exemption of the heavy industry scenario is less than that observed for the other two scenarios. After the year 2007 the decline starts accelerating.

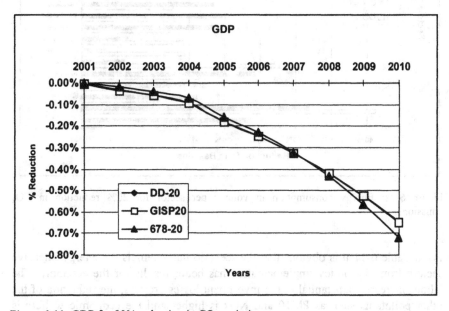

Figure 6-11: GDP for 20% reduction in CO_2 emissions.

The energy intensive sectors that were obliged to reduce their emissions in the basic double dividend scenario, are now free to emit as much as it is necessary in order to minimise their costs and maximise their profits. In 678-20 scenario the pure energy sectors such as the oil, gas and electricity sectors along with the other service and equipment goods industries are reducing their energy consumption more, in comparison to the original double dividend scenario, in order to compensate the reduction that it should have been done from the energy intensive sectors. The energy intensive sectors do not have to substitute energy with other production factors since they do not face a higher price in energy. The small reduction that is observed can be explained from the fact that after a double dividend policy application they face a smaller cost of labour so they are able to substitute energy with the less expensive labour. The sectoral results concerning the energy consumption are shown in Figure 6-12 below.

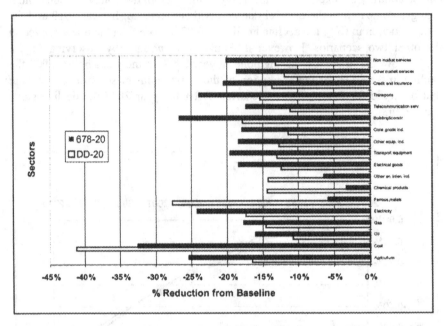

Figure 6-12: Energy consumption in volume per sector for 20% reduction in CO_2 emissions.

From Table 6-11 it is obvious that the scenario that exempts the energy intensive sectors from the policy implementation has better results for the economy. The labour increases substantially, the investment losses are less, the reduction of the other pollutants such as PM10 and NOx is higher and the economic welfare is increased. However it should be mentioned that the marginal abatement cost for achieving a 20% reduction in CO_2 emissions, leaving the energy sectors out is considerably more in the 678-20 scenario compared to the other two. In year 2010 the marginal abatement cost is 240 ECU per ton of carbon for the DD20 and the GISP20 scenarios and it almost doubles (438 ECU/ton of carbon), when the energy intensive industries are not obliged to reduce their emissions. Thus, the economy should double its effort in order to achieve the same emission reduction target. The marginal abatement cost for the European Union is shown in Figure 6-13 below.

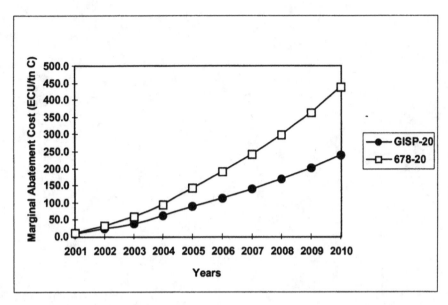

Figure 6-13: Marginal abatement cost.

7 The Role of Energy Saving Investment

7.1 Model Specification

In this simulation the model is modified to incorporate the additional possibility of investing in energy-saving technology. In this context economic agents are assumed to have the possibility to increase the productivity of energy (i.e. consume less energy inputs per unit of output) by investing some resources to accumulate an energy-saving technology stock.

The decision whether and how much to invest on energy saving technology is, in this model variant, an *endogenous decision of the economic agents*.

Producers in the standard version of the GEM-E3 model demand the cost-minimising mix of factors needed for production (capital, labour, and energy and intermediate goods). Based on the demand they face, their existing production capacity and their expectation about the future, they also invest to increase their productive capital stock. In the new model variant GEM-E3 ver. 2.a, there is an additional production factor, the stock of energy-saving technology, that serves to reduce the amount of energy needed to produce the same output. The term "energy-savings" is used to indicate all technologies that may improve energy productivity.

The specification follows the Grossman/Helpman "expanding product variety" approach, as this draws from the Dixit/Stiglitz conception of "product variety". While the methodology is standard in theoretical literature, this is one of the first attempts to incorporate endogenous technical progress in a full-scale applied CGE model.

114

Producers compare the benefit of having one more unit of additional energy saving technology[28] (that will induce lower energy costs for all subsequent time periods) to the cost of acquiring this additional unit. Through their cost-minimising behaviour, they decide (as the solution of their inter-temporal problem of allocating their resources), together with the demand for production factors and for productive investments, the optimal investment in energy saving technology. This investment does not affect their productive capital stock, but is an additional demand for goods and related services (which for example may be equipment goods and the services needed for installation or operation of the new equipment).

In addition, in the new model variant GEM-E3 Ver 2.a, consumers have the possibility to choose among two classes of durable goods: "ordinary" ones, and more energy efficient ones, say "advanced technology", whose acquisition price is however higher. A trade-off between higher acquisition costs for durable and lower energy (operation) costs, is thus introduced in the decision making of the consumer.

The investments related to energy saving accumulate to form an energy-savings capacity (stock) that of course will have permanent effects on energy productivity. Similarly, the choice of more efficient durable goods will also have permanent effects on the use of energy in households. The relationship between the energy-saving stock and its effectiveness for energy savings exhibits decreasing marginal returns (saturation effect). For example if a constant investment on energy savings is accumulated per year, the resulting increase to the stock is increasing at decreasing rates, as illustrated in the Figure 7-1 below.

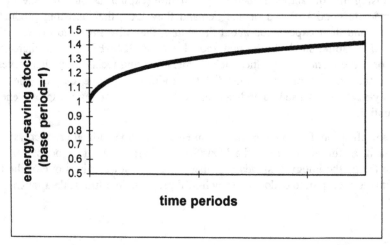

Figure 7-1: Energy-saving stock under a constant investment in energy-saving technology.

---

[28] Or, equivalently, increase the "quality" of the energy saving stock at their disposal.

The basic algebraic formulation links the energy-saving expenditure to the stock of energy-saving capital and the increase of that stock i.e. its marginal effectiveness. For example, if the energy use costs increase, the firm may quickly invest on energy saving as long as the accumulated energy-saving capital remains low. Beyond a certain level of saturation, the firm may decrease further investment on energy-saving, as its marginal effectiveness decreases and the corresponding expenditure that is necessary to maintain a certain level of effectiveness increases non linearly. This is illustrated in the following figure.

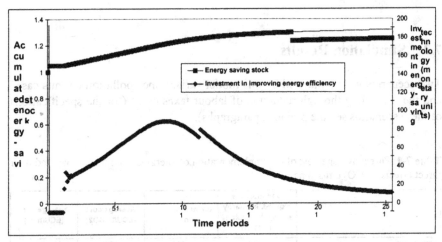

Figure 7-2: Energy saving investment and stock under a continuous 3% increase of energy prices every year.

The following remarks must be added.

The energy-saving stock acts cumulatively, i.e. energy-saving technology accumulated by year t continues to be available in all subsequent years at no additional cost. A stock of energy-saving technology once acquired cannot be scrapped.

The energy saving technology is depletable, i.e. there is an upper limit to the amount of energy productivity gains that can be achieved, a limit which depends on the particular economic agent and country (see Figure 7-2 above).

The desired amount of energy-saving technology depends mainly on the relative price of energy compared to the cost of acquiring it (which in its turn depends on the stock of energy-saving technology already accumulated, see Figure 7-2 above) and the share of energy inputs in production costs.

Through this mechanism, the model incorporates an important dynamic mechanism, which is related to the depletable character of the energy savings potential. This mechanism involves a trade-off over time, as the firms (or consumers), may prefer to invest on energy savings (allocating less expenditure to consumption or investment) expecting to then use energy more productively.

The model variant is applied to the analysis of the emission targets, under the same terms of the other exercises mentioned in the previous sections. To better appreciate the dynamics of the results the model simulation goes up to 2020, with the additional assumption that after 2010 there is no additional environmental policy implemented within the EU above what is needed to keep emissions stable to their 2010 levels.

7.2 Simulation Results

Two scenarios were simulated by this model variant: the "pollution permits case" and the "recycling through reduction of labour taxes case" (for the specifications of these scenarios see the previous paragraphs).

Table 7-1: Direct average cost of the implementation of energy-saving technology and their direct effects on CO_2 emissions.

	Commulative expend. in energy saving technology (in MECU)	Abated Emissions (ktn CO2)	Average cost (ECU/tn CO2)	Average cost (ECU/tn C)
Agriculture	1427	113626	12.6	46.1
Electricity	9317	388908	24.0	87.8
Ferrous, non ferrous ore and metals	4806	326345	14.7	54.0
Chemical products	2793	138657	20.1	73.9
Other energy intensive industries	1884	96746	19.5	71.4
Electrical goods	121	14536	8.3	30.4
Transport equipment	239	24776	9.6	35.4
Other equipment goods industries	188	20397	9.2	33.8
Consumer goods industries	1293	105602	12.2	44.9
Building and constraction	1185	97445	12.2	44.6
Telecommunication services	77	10014	7.7	28.1
Trasports	3932	205483	19.1	70.2
Credit and insurance	178	19108	9.3	34.1
Other martket services	3891	259567	15.0	55.0
Non-market services	1875	164699	11.4	41.7
Totals	35353	2034077	17.4	63.7

The main additional mechanism operating in GEM-E3 ver. 2.0a is as follows: the increase in energy prices, increase the long-term profitability of investing in energy saving technology. Thus, compared to the main model simulations (presented in the previous chapters), an amount of resources that was previously allocated to consumption and productive investments in now dedicated to energy-saving investments. This has two important implications: on the one hand it leads to less dependency on imports (as energy is mostly an imported good) and, on the other hand, it boosts domestic demand in those specific directions that deliver the energy-saving technology, such as the equipment goods industry and related services. This has a positive impact both on GDP and on investments. As the potential for energy savings is gradually depleted, energy-saving investments become lower, although the gains, in energy efficiency obtained from previous investment remain. During the transition period (up to 2010), employment is higher than in the main model results (as the result of increased activity) but then gradually it is reduced at a level lower than in the main model simulation. GDP losses are much alleviated, also due to terms-of-trade effects that are important during the transitory period.

Table 7-1 shows the cumulative expenditure in energy-saving technology and the corresponding gains in the productivity of energy.

Employment gains are again effected. In the transitory period (up to 2010) employment is even higher than in the main model simulation (1675 thousand employed persons as opposed to 1460 with the main model), but is gradually reduced to 1342 thousand by year 2020.

GDP losses are much alleviated both in the short and the longer term. The decrease in GDP is reduced by half, up to 2010 (-0.33% as opposed to -0.65% in the main model simulation) mainly because of the demand for energy-saving investments and after this point by terms-of-trade effects related to the fact that less energy imports are needed. In the first years of the environmental policy GDP is even slightly increasing.

Terms-of-trade gains from the rest-of-the-world again explain part of the positive results.

An important dynamic issue not covered in this model version, is the possibility that the accumulation of energy saving capital may shift the possibility frontier of energy saving investments so that additional productivity gains may be possible. In such a case the longer-term path of GDP and welfare would probably tend to return to the values of the baseline scenario.

Table 7-2: Macroeconomic aggregates with energy saving investment for the double dividend case and for 20% reduction in CO_2 emissions (EU-14 in 2010).

	2001	2006	2010	2014	2020
Gross Domestic Product	0.01%	-0.06%	-0.33%	-0.34%	-0.36%
Employment*	163	785	1675	1411	1342
Private Ivestment	-0.02%	-0.15%	-0.43%	-0.39%	-0.41%
Investments in energy-saving technology**	0.29%	0.90%	1.50%	0.21%	0.09%
Private Consumption	0.14%	0.68%	1.24%	1.21%	1.15%
Domestic Demand	-0.03%	-0.35%	-1.00%	-1.16%	-1.20%
Exports in volume	-0.51%	-2.52%	-5.45%	-4.41%	-4.23%
Imports in volume	-0.03%	-0.61%	-1.60%	-2.30%	-2.37%
Intra trade in the EU	-0.53%	-2.59%	-5.59%	-4.50%	-4.31%
Energy consumption in volume	-1.10%	-6.45%	-14.48%	-14.51%	-14.50%
Consumers' price index	0.49%	2.53%	5.76%	4.91%	4.71%
GDP deflator in factor prices	0.17%	0.60%	1.05%	0.35%	0.27%
Real wage rate	0.05%	0.39%	0.85%	0.98%	0.88%
Tax revenues as % of GDP***	0.28%	1.63%	3.78%	3.61%	3.52%
Current account as % of GDP***	-0.05%	-0.17%	-0.33%	-0.17%	-0.15%
Marginal abatement cost (ECU'85/tn C)	13.6	86.9	232.9	220.4	214.6
Total atmospheric emissions					
CO2	-1.98%	-9.97%	-19.97%	-19.97%	-19.97%
NOX	-1.46%	-7.61%	-16.27%	-16.33%	-16.27%
SO2	-3.13%	-15.52%	-29.19%	-29.00%	-28.97%
VOC	-0.74%	-4.01%	-10.03%	-10.17%	-10.07%
PM	-3.58%	-17.25%	-31.97%	-31.77%	-31.70%

* in thousand employed persons
** as a percent of private investments
*** absolute difference from baseline
**** as percent of GDP (last simulation year)

8 Internalisation of Externalities

8.1 Introduction

The internalisation of external effects is an important policy guideline in the Energy and Environmental policy at EU level. The objective of this chapter is to evaluate the macroeconomic and welfare impacts of a policy aiming at applying this guideline to the externalities generated by energy consumption. It will also contribute to a *comparison of an integrated policy approach towards air compared to a policy addressing only the global warming issue.* In most sectors these emissions are one of the main sources of external cost. A policy scenario will be defined in which an environmental cost is applied on the emissions in proportion to the damage generated by these emissions.

The second section will briefly recall the methodology of zoned emission taxes used in this paper. The third section will discuss the valuation assumptions used to construct the marginal damage data and the emission taxes. In the fourth paragraph the construction of the emission tax scenario is discussed. Results of this scenario are analysed in the fifth paragraph.

8.2 Methodology

8.2.1 The Principle of a Pigouvian Tax

For one pollutant and one location of emissions, the effects of the introduction of a Pigouvian tax can be shown easily by the following basic diagram.

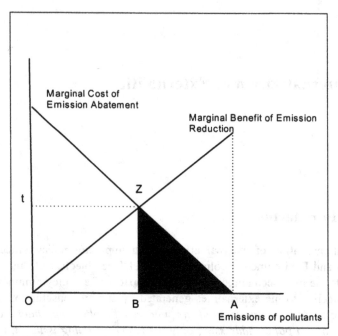

Figure 8-1: The effects of a Pigouvian tax.

This diagram shows two lines starting from point A on the X-axis. This point represents the initial emissions (before any specific environmental policy). Line AZ represents the marginal cost of emission abatement (MAC). This line by definition ranks all available possibilities to reduce emissions as a function of their costs. It is upward sloping from right to left because the cost of emission reduction increases as emissions are reduced. The second line (MB) represents the savings of environmental damage when emissions are reduced in the direction A to O. The marginal damage curve represents the sum, over all victims of pollution, of the damages avoided. This curve can slope as well upwards from right to left.

The optimal level of emission reduction is achieved when the sum of the total damages avoided minus the total abatement costs is maximised. In the absence of corner solutions, this implies point Z and an emission level B where the marginal cost of emission reduction equal t, which is also the marginal damage saved.

Emission taxes and emission standards can be used to make polluters choose level Z. Standards can be used if the marginal abatement cost is more or less known for each polluter. In order to select point Z one needs to know also the shape of the marginal benefit curve. An emission tax equal to Ot can achieve the same optimal level of abatement B much easier. Knowledge of point Z is sufficient. All polluters have to pay the marginal emission tax and this allows decentralising the choice of emission reduction levels to individual agents. All polluters with a marginal cost lower than the emission tax will find it profitable for them to avoid taxes by making more abatement efforts. This implies that the emission reduction will be distributed in a cost-effective way in the case of emission taxes.

In this case study, information over the marginal benefits of emission reduction will be limited to one value: one assumes that the MB is a horizontal curve. This is the ideal case for applying an emission tax, as the (constant) MB is the only information needed to determine the appropriate emission tax. This implies (Figure 8-2) that the implementations of an emission tax generates a net benefit equal to the shaded area: the difference between the total emission damages saved and the total abatement costs of reaching level B.

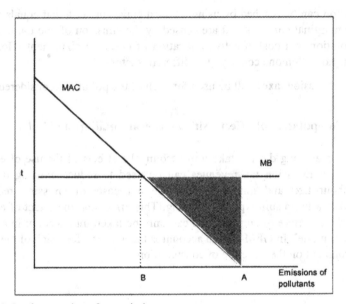

Figure 8-2: Implementation of an emission tax.

8.2.2 A Zone Based Pigouvian Tax in a European Context

It is well known that for most forms of air pollution, damage is of a transboundary nature. Take Belgium as an extreme example: for a pollutant like VOC, more than 90 % of the marginal damage originating in Belgium is located outside its borders. This implies that for every country of origin the sum of the damages caused in all other countries caused by these emissions, will be different and that a different emission tax has to be used for each country where emissions originate. More specifically, in every country i, the marginal damage per ton of pollutant deposited MD_i is taken constant, from every ton of pollution emitted in country j a share s_{ij} is deposited in country i, the total abatement cost of emissions in country j is given by TAC_j. Let the initial emission level be equal to E^o_j

The environmental problem is to minimise

$$\sum_i MD_i {}^* \sum_j s_{ij} {}^* E_j + \sum_j TAC_j {}^* (E^o_j - E_j)$$

In the absence of a corner solution the optimal solution requires that

$$\sum_i MD_i {}^* s_{ij} = MAC_j (E_j) = t_j$$

This solution can be reached by using a zoned emission tax t_j that equals the sum of all the marginal damages that are caused by the emission of one ton in country j. This solution is a cost effective allocation of emission abatement efforts over countries and within one country over different sectors.

The zoned emission taxes will be used for each of the pollutants considered.

8.2.3   Computation of Effects with a General Equilibrium Model

The above reasoning did not take into account the effects of the use of emission tax revenues. Emission tax revenues can be used to reduce existing distorting taxes (labour tax) and this can lower the net costs of emission reductions compared to a lump sum type of recycling. This effect and the effect of emission taxes on the structure of consumption can only be taken into account in a general equilibrium model. In GEM-E3, no account is taken of the feedback of the state of the environment on the structure of consumption.

8.3   The Externalities and their Damage Valuation

Three environmental problems (i) global warming, (ii) problems related to the deposition of acidifying emissions and (iii) ambient air quality problems linked to acidifying emissions and ozone concentration are considered, i.e. the externalities of the air pollution type which are mainly generated by energy-related emissions of NO_x, SO_2, VOC and particulates (no solid waste, noise or accidents issues are examined). NO_x SO_2 and particulates are almost exclusively generated by combustion processes, whereas VOC's are only partly generated by energy using activities like refineries, combustion of motor fuels; other important sources of VOC's are the use of solvents in the metal industry and in chemical products. These non energy-related emissions are not taken into account here.

The damage occurs when primary (e.g. SO_2) or secondary (e.g. $SO_4^=$) pollutants are deposited on a receptor (e.g. in the lungs, on a building) and, ideally, one should relate the physical damage per receptor to deposition per receptor. In practice, dose-response functions are related to (i) ambient concentration to which a receptor is submitted, (ii) wet or dry deposition on a receptor or (iii) 'after deposition' parameters (e.g. the PH of a lake which is related to acid rain). Following the 'damage or dose-response function approach', the incremental physical damage per country is given as a function of the change in deposition/concentration (acidifying components or ozone concentration in the model).

The damages categories considered here are:

1. damage to public health (morbidity and mortality, chronic morbidity, but no occupational health effect);

2. damage to the territorial ecosystem (agriculture and forests) and to materials, treated in a very aggregated way;

3. global warming.

The ExternE project retains, as principal sources of health damages from air pollution, particulates[29], resulting from direct emission of particulates or due to the formation of sulphates (from SO_2) nitrates (from NO_x), and ozone. They retain also the direct effect of SO_2 but not the direct impact of NO_x because it is likely to be small. The assessment of health impacts is based on a selection of exposure-response functions from epidemiological studies on the health effects of ambient air pollution (both for Europe and for the US). These public health impacts are far more important than the damage to agriculture, forests and materials.

Damage on agriculture and forest is done by foliar uptake or by acid deposition on the soil. ExternE has examined in detail the assessment of the direct impact of SO_2 and O_3 on yield of a limited number of crops (rye, oats, barley, peas, beans and wheat) through the use of exposure-response functions, on the increased liming requirement to compensate for acid deposition and on the reduced nitrogenous fertiliser requirements through deposition of oxidised N. They made some tentative estimations of the forest damage, more with a methodological objective than with the objective to give complete results for the assessment of the forest damage. Discoloration, material loss and structural failure are the main impact categories for most materials (mainly through corrosion), which result from interactions with acidifying substances like SO_2 and NO_x, particulates and ozone. The impact is highly dependent on the material in question: buildings, textile, paper, etc. Because of the great uncertainty in the dose response functions, the

---

[29] PM10, i.e. particulates of less than 10 $\mu g/m^3$ aerodynamic diameter, is taken as the relevant index of ambient particulate concentrations.

aggregation level of GEM-E3 and their likely insignificance[30], an average damage cost per person was used for this category of damages in GEM-E3, based on ExternE results.

The economic valuation of the damage should be based on the willingness-to-pay concept. For market-goods, the valuation can be performed using market prices. When impacts occur in non market goods, two broad approaches have been developed to value the damages. The first one, the contingent valuation approach, involves asking people open- or closed-ended questions for their willingness-to-pay in response to hypothetical scenarios. The second one, the indirect approach, seeks to uncover values for the non marketed goods by examining market or other types of behaviour that are related to the environment as substitutes or complements. Depending on the type of damages, both approaches were used in ExternE. The following figures (Table 8-1), derived from the results in ExternE, were used in this study.

Table 8-1: Damage from an increase in air pollution (10^3 ECU per 1000 persons).

Public Health	from an increase of one $\mu g/m^3$ of PM10* concentration	0.005358
	from an increase of one $\mu g/m^3$ of nitrates concentration	0.005358
	from an increase of one $\mu g/m^3$ of sulphates concentration	0.008351
	from an increase of one $\mu g/m^3$ of SO_2 concentration	0.000693
	from an increase of one ppb of ozone concentration	0.00243
Other	from an increase of one $\mu g/m^3$ of sulphates concentration	0.0028
	from an increase of one $\mu g/m^3$ of nitrates concentration	0.0018

*Particulate Materials

The transboundary nature of pollutants requires specifying the transport of SO_2, NO_x, VOC and particulates emissions between countries. In the case of tropospheric ozone (a secondary pollutant), there is another difficulty as tropospheric ozone concentration is a non linear function of VOC and NO_x emissions, the two ozone precursors. Theoretically, the concentration/deposition of a pollutant in a grid is a function of the total anthropogenic emissions before

---

[30] The first results from ExternE showed that they were relatively less important than public health impact.

time t, some background concentration in every country[31], and other parameters, such as meteorological conditions and chemical reactions of pollutants. The transport of pollutants and their marginal effect are approximated using linear transfer coefficients. These reflect the effect of the emissions in each country on the deposition/concentration of a pollutant in another country. At this stage no distinction is made between the sectoral origin of the emissions.

For global warming, the global atmospheric concentration matters and it is only a function of the total anthropogenic emission of greenhouse gases.

8.4 The Internalisation Scenario

The policy of internalisation aims at imposing environmental taxes on production and consumption corresponding to the total environmental damage (in monetary terms) that is linked to these economic activities (both on the national territory and abroad). Consumers and producers will then take this extra cost into account in their decision on pollution abatement, production and consumption.

This extra cost is implemented in this scenario as a tax per unit of emission, the same for all consumers and producers in one country. This tax is added to the existing taxes as almost no environmental taxes exist as yet. For global warming, the treatment is slightly different. Estimation of the damage from CO_2 emissions varies between 1 and 30ECU/tCO_2. Because of this large range in order to allow for comparability between this scenario and the other scenarios with GEM-E3 in this volume, an EU-wide tax for CO_2 is computed endogenously in the model to reach the EU target of 10% reduction in 2010 compared to the level of 1990 emissions. Assuming an emission growth of 8% for the EU, this implies an actual reduction effort for the EU of 18%. The policy is gradually imposed from 2001 onwards and it is assumed that no other environmental policies are in place.

The emission tax is constructed by combining the damage estimates of Table 8-1 with transfer matrices from EMEP and from ExternE. The wind directions and the population density are the principal explanations for the differences in emission taxes among countries. Countries like Greece or Spain have relatively low damaging impacts on the rest of the EC consequences and have lower emissions taxes than countries like Germany or Belgium where wind currently leads the polluting emissions into densely populated zones of the European Union.

---

[31] Resulting from natural emissions and emissions from geographic parts that are not included in the country set.

Table 8-2: Emission Tax (ECU/kg).

	NOx	SO$_2$	VOC	PM
Austria	2.02	1.00	0.07	3.38
Belgium	5.26	3.10	0.26	4.55
Germany	4.21	2.22	0.22	3.80
Denmark	2.39	1.03	0.07	2.08
Finland	1.07	0.36	0.01	1.06
France	4.11	1.79	0.12	3.43
Greece	0.82	0.32	0.03	1.57
Ireland	3.06	0.98	0.09	1.85
Italy	2.52	1.01	0.10	2.95
Netherlands	4.72	3.07	0.24	4.11
Portugal	1.63	0.62	0.08	1.65
Spain	2.35	0.66	0.05	2.43
Sweden	0.99	0.45	0.02	1.10
United Kingdom	3.54	1.94	0.21	3.05

Regarding the recycling strategy, it is assumed that the additional income is used by the public sector to reduce the employer's social security contributions uniformly in all sectors. In a previous chapter the impact of alternatives for recycling strategies were compared.

This scenario will be compared with the "double dividend" scenario presented in previous chapter where only the CO_2 emission target is imposed and the tax being computed endogenously to reach this target. The differences between the scenarios allow the evaluation of the benefits from an integrated approach concerning air pollution.

Though a CO_2 tax can be relatively easy to implement because these emissions are directly linked to the energy consumption, this is not true for the other pollutants for which the monitoring of emissions can be sometimes either very expensive or not possible.

8.5   Simulation Results

The results are given for 2010, when the emission target is fully implemented. As expected the reduction in emissions is much higher in the internalisation scenario (except for the CO_2 emissions because of the scenario construction). It is also of interest that the marginal cost of CO_2 reduction (at the end of the reduction period) is nearly halved.

Table 8-3: Impact on EU-wide emissions and CO_2 reduction cost.

	Internalisation scenario	CO_2 tax scenario
Nox	-31%	-14%
SO_2	-34%	-26%
VOC	-14%	-9%
PM	-30%	-28%
CO_2	-18%	-18%
marginal cost of CO_2	27 ECU85/tn CO_2	49 ECU85/tn CO_2

The total user energy cost (including the emission taxes and the marginal abatement cost) is not too different in both scenarios: the higher tax on other pollutants and the abatement cost are compensated by the lower CO_2 tax. There is however a sectoral differentiation: in sectors where there are large abatement possibilities, the increase in the energy cost is smaller in the internalisation scenario compared to the double dividend scenario whereas the reverse is observed in sectors where the abatement possibilities are small.

The impact at a macroeconomic level of both scenarios is rather similar. The tax revenues are slightly higher in the double dividend scenario because of the abatement possibilities for the other pollutants, explaining the higher impact on employment. The abatement costs are 'recycled' as intermediate demand (partly domestic, partly imported) instead of lowering the social security rate.

Table 8-4: Macro-economic impacts for EU-14.

	Internalisation scenario	CO_2 tax scenario
Gross Domestic Products	- 0.2%	-0.0%
Employment*	1049	1178
Private Investment	-0.5%	- 0.6%
Private Consumption	0.2%	0.5%
Exports in volume	-2.5%	-2.5%
Imports in volume	-2.7%	-3.2%
Energy Consumption	-12.1%	-12.1%
Tax Revenues as % of GDP	2.9%	3.0%
GDP deflator in factor prices	-1.4%	-1.7%
Current account as % of GDP	0.1%	0.2%
Economic welfare**	0.1%	0.1%
Economic and environmental welfare**	0.4%	0.3%

* in thousand employed persons
** as percentage of GDP

Both scenarios obtain a positive, though small, economic welfare effect, expressed in terms of GDP. The gain in welfare through the increase in consumption compensates the loss due to the decrease of leisure. In terms of GDP growth and foreign trade the impacts remain rather similar in both scenarios. However considering the total welfare, i.e. including the environmental benefits, the internalisation scenario has a greater positive effect because the reduction in total emissions is higher.

Private consumption is slightly higher in the CO_2 scenario because households benefit from the higher impact on employment and real wage in that scenario. The impact on investment remains similar, the higher substitution effect towards labour being compensated by a higher demand effect.

Table 8-5: Impact on private consumption and private investment.

	Consumption		Investment	
	Internalisation	CO_2	Internalisation	CO_2
Austria	0.5%	0.6%	-0.4%	-0.4%
Belgium	0.1%	0.7%	-0.7%	-0.6%
Germany	-0.1%	0.3%	-0.5%	-0.5%
Denmark	0.5%	0.5%	-0.4%	-0.5%
Finland	0.2%	0.3%	-0.3%	-0.4%
France	-0.4%	-0.1%	-0.6%	-0.5%
Greece	-0.6%	-0.8%	-0.8%	-1.1%
Ireland	0.1%	0.1%	-0.8%	-1.0%
Italy	-0.5%	-0.5%	-0.5%	-0.6%
Netherlands	0.0%	0.2%	-0.4%	-0.4%
Portugal	-0.5%	-0.6%	-0.9%	-1.1%
Spain	0.1%	0.2%	-0.8%	-0.8%
Sweden	0.3%	0.9%	-0.3%	-0.4%
United Kingdom	-0.3%	0.2%	-0.7%	-0.6%

Exports decrease in both scenarios because of the loss in competitiveness, but this negative impact is partly compensated by the higher export prices. The imports also decrease because of the sharp drop on the energy consumption. Therefore no deterioration of the current account in terms of GDP is observed. This result depends partly on the possibility of shifting some of the burden of the tax reform to the rest of the world and is therefore sensible to the assumed export and import price elasticities.

Table 8-6: Impacts on foreign trade.

	Exports		Imports	
	Internalisation	CO_2	Internalisation	CO_2
Austria	-3.0%	-3.1%	-1.8%	-2.5%
Belgium	-3.4%	-2.9%	-2.7%	-2.3%
Germany	-3.3%	-2.8%	-2.2%	-2.3%
Denmark	-4.0%	-3.9%	-4.4%	-4.7%
Finland	-2.6%	-2.9%	-0.5%	-1.2%
France	-2.7%	-2.0%	-3.5%	-3.3%
Greece	-2.5%	-3.2%	-3.2%	-4.6%
Ireland	-3.4%	-3.4%	-0.6%	-1.0%
Italy	-2.0%	-1.5%	-4.5%	-4.7%
Netherlands	-1.8%	-1.4%	-0.9%	-0.9%
Portugal	-1.4%	-1.2%	-1.9%	-2.2%
Spain	-3.7%	-3.5%	-5.3%	-5.7%
Sweden	-2.4%	-3.1%	-2.7%	-3.6%
United Kingdom	-1.9%	-1.6%	-1.4%	-1.4%

9 Burden Sharing

9.1 Introduction

The recent developments in climate change policy reveal once more the importance of the burden sharing issue in fleshing out an internationally binding commitment to reduce greenhouse gas emissions. The heterogeneity of interests involved stems from national and/or sectoral differences in wealth, the current stage of development, the energy intensity of production and/or consumption, the resource allocation, and so on. The debate on how to allocate the burden entailed by climate protection revolves around international and regional commitment efforts. For the climate change policy of the European Union (EU) this issue arises at both levels.

At both levels, regionally and internationally, pollution permits are often considered to be the most appropriate policy instrument for an efficient implementation of reduction targets for global pollutants. While the international trade of pollution permits equalises marginal costs of emission reduction across countries and, therefore, provides an efficient solution, the initial allocation of permits enables various opportunities to tackle the equity aspect.

This chapter deals with both efficiency and equity, whereas the main emphasis lies on the latter. The objective is to review and analyse the potential impacts of different equity rules applied to the EU-internal burden sharing.

The first part of the chapter (Sections 9.2 and 9.3) deals with the principle concepts and alternatives of the equity issue from a methodological point of view and refers to the initial allocation of pollution rights in particular. The second part of the paper applies a selection of equity rules to a given EU reduction target (10% reduction of CO_2 by 2010 based on 1990 emissions) (Section 9.4) and assesses the potential impacts with respect to welfare and other macro-economic aggregates on both EU-wide and national levels (Section 9.5). All simulations are undertaken with the computable general equilibrium model GEM-E3. The final section (Section 9.6) draws some conclusions.

9.2   Equity and Efficiency

An effective protection of global common goods requires an international agreement whereby countries commit themselves to emission reductions on a more or less voluntary basis. As countries will not be willing to cooperate if they feel that the international sharing of economic costs of the global warming policy is unfair, international agreements should be based on equity criteria that are accepted by all negotiators. In general, international equity issues arise in view of substantial differences between countries, e.g. in terms of population, wealth and consumption, purchasing power, emissions of greenhouse gases in the past, present and future, or vulnerability to climate change (Banuri et al. 1996, p. 91). The choice of the equity rule has significant consequences for the distribution of the costs of environmental protection amongst countries.

No political consensus on who should pay for abatement of CO_2 at the international level has been developed yet. As equity issues are not a topic of economic theory, equity principles have to be based on ethical and normative judgements. However, studying the potential impacts of different burden sharing mechanisms might give some insights that could contribute significantly to reach an international consensus.

As long as the theoretical conditions for an efficient allocation prevail, economic efficiency, defined in terms of Pareto-optimality, is neutral with respect to distributional issues. The economically efficient allocation of emission reductions is independent of how the total costs of these reductions are shared between countries, or which equity rule is applied. Empirical analysis supports the theoretical finding of the independency of the Pareto-optimal emissions trajectory from the initial allocation of emission rights.

Manne and Richels (1995, p. 23), for instance, note that "the rules for the allocation of emission rights imply that there will be wealth transfers, but empirically we have found that the general equilibrium effects of these transfers are too small to influence the overall level of emissions. In this sense, our results are consistent with the Coase Theorem - the proposition that wealth transfer effects are too small to influence the Pareto-optimal level of provision of a public good (the mean global temperature)".

Conflicts between efficiency and equity principles in global warming policy arise only in the absence of international lump-sum transfers. According to Tinbergen (1952, pp. 27ff.), each policy that is supposed to address both efficiency and equity needs two policy instruments: one instrument for dealing with allocable objectives, e.g. a tax or permit system, and a second for implementing equity issues, e.g. compensation payments or transfers. In this sense, transfers are not only interpreted as financial transfers, e.g. redistribution of tax revenues, but can also be in terms of an initial free-of-charge allocation of emission permits.

According to this, the IPCC report (Banuri et al. 1996) makes a distinction between proposals for distributing abatement costs including international transfers and proposals for fair distribution of emissions separated from any substantive financial transfers.

Under a non cooperative approach without compensation payments, the initial distribution of emission constraints determines simultaneously both the allocation of emission reductions as well as the distribution of costs. Given the assumption that the international authority has full information concerning national abatement cost functions, a cost-efficient international allocation of emission reductions can be implemented. But even if efficiency of resource allocation is ensured, the resulting distribution of burden sharing might be perceived as unfair as it "does not address the tendency toward national self-interest, the ability of nations to pay the cost, or international political balance" (Rose 1990, p. 927). On the other hand, an equitable allocation of emission constraints is very likely to be inefficient from a global point of view.

Under a cooperative approach, an overall emission reduction target, which is defined for the whole group of the contractual partners, is implemented by a common economic instrument, such as an international CO_2 tax or an international tradable permit system. Here, efficiency and equity aspects can be addressed separately and treated as complementary principles. Whereas the allocation/price mechanism of the economic instrument leads to an efficient allocation of emission reductions, equity concepts are used to determine how the tax revenue is distributed or how the permits are allocated initially across countries.

Even if the literature provides different classifications of principles for burden sharing, the "examination of international equity issues is still in its infancy" (Banuri et al. 1996, p. 85). Banuri et al. (1996, p. 104), for example, distinguish between procedural and consequential equity; the first has to do with participation, process and treatment before the law, the second refers to the outcome of decisions, i.e. the distribution of burdens and benefits. The latter is split in five categories: parity (all claimants receive equal shares of burdens or benefits), proportionality (distribution of burdens or benefits in proportion to the contributions of claimants), priority (those with the greatest needs should be put first), classical utilitarism (distribution in order to achieve the greatest good for the greatest number), and Rawlsian distributive justice (equal distribution unless an unequal distribution operates to the benefit of the least advantaged).

In the following, different mechanisms for burden sharing, mainly provided by Rose and Stevens (1996), and Rose (1990), are presented and discussed regarding their implications for the international distribution of costs and benefits of global warming policy. Table 9-1 summarises the criteria and translates them into operational rules for global warming policy in general and tradable permits in particular. Keeping in mind that economic efficiency and equity issues can be separated in economic theory, the listed principles of burden sharing can also be used to determine reimbursement rules in the case of a CO_2 tax.

First of all, Rose and Stevens classify the equity criteria according to whether they are defined in terms of the initial allocation of emission rights ('allocation-based') or in terms of traditional welfare economics ('outcome-based').

Allocation-based rules are related directly to the distribution of emission rights or to the distribution of gross abatement costs. Within a model framework, allocation-based rules are reflected by each country's initial endowment of emission rights according to the applied equity criterion. The distribution of emission rights may represent an initial stock of tradable permits in case of a cooperative implementation of the global reduction target (i.e. subsequent trading of permits between countries is allowed) or the final allocation of national emission levels in case of a non cooperative solution (i.e. countries implement given reduction targets independently).

In contrast to allocation-based rules, outcome-based rules take into account the incidence of costs and benefits, i.e. the net welfare change due to global warming policy. Outcome-based rules are defined in terms of the relation of net welfare to sub-criteria such as GDP or GDP per capita. Modelling this rule is more complex as for each country one has to set the net welfare level, the equity rule being simulated demands. Then, the model is solved either for the corresponding initial permit allocation, taking into account the international trading of emission permits (in case of a cooperative implementation of the global reduction target), or for the corresponding final allocation of national emission levels (in case of a non cooperative implementation of the global reduction target).

Table 9-1: International equity criteria for sharing the costs and benefits of a global warming policy.

Criterion	Basic Definition	General Operational Rule	Operational Rule for CO_2-Permits
Outcome-Based Equity Criteria			
Horizontal	All nations should be treated equally	Equalise net welfare change across nations (net cost of abatement as proportion of GDP equal for each nation)*	Distribute permits to equalise net welfare change (net cost of abatement as proportion of GDP equal for each nation)*
Vertical	Welfare changes should vary inversely with national economic well-being	Progressively share net welfare change across nations (net cost proportions inversely correlated with per capita GDP)*	Progressively distribute permits (net cost proportions inversely correlated with per capita GDP)*
Compensation	No nation should be made worse off	Compensate net losing nations	Distribute permits so that no nation suffers a net loss of welfare
Rawls's Maximin	Maximisation of the welfare of the worst-off nations	Maximise the net benefit to the poorest nations	Distribute largest proportion of net welfare change to poorest nations
Allocation-Based Equity Criteria			
Ability to Pay	Mitigation costs should vary inversely with national economic well-being	Equalise abatement costs across nations (gross abatement costs as proportion of GDP equal for each unit)**	Distribute permits to equalise abatement costs (gross costs of abatement as proportion of GDP equal for each nation)**

136

Table 9-1: Continued.

Egalitarian	All people have an equal right to pollute and to be protected from pollution	Equalise per capita emissions across countries (in contemporary or historical form)	Distribute permits in proportion to population or historical responsibilities
Sovereignty	All nations have an equal right to pollute and to be protected from pollution	Allocation of emission reductions in a proportional manner across all nations, i.e. equal percentage emission reductions	Distribute permits in proportion to emissions
Market Justice	The market is fair	Make greater use of markets	Distribute permits to highest bidder
Consensus	The international negotiation process is fair	Seek a political solution promoting stability	Distribute permits in a manner that satisfies the (power weighted) majority of nations

*Net costs equal to the sum of: mitigation benefits - abatement costs + permit sales revenues - permit purchase costs.

** Gross cost refers to abatement costs only.

Source: Rose and Stevens (1996, p. 3), modified.

9.3   Operationalisation of Equity Rules

9.3.1   Outcome Based Equity Criteria

Horizontal equity calls for an equal treatment of all nations in terms of welfare outcome. Rose and Stevens (1996, p. 11) distinguish two different operational versions of the horizontal equity concept.

- The share of net welfare to GDP is equalised across nations. This implies that countries with a relatively larger share of global GDP receive a relatively greater share of global net welfare.

- The share of net welfare to population is equalised across countries. Consequently, the largest share of global net benefits is distributed to those countries with the largest share of global population.

A third operational rule of horizontal equity can be defined as follows.

- The share of net welfare to GDP per capita is equalised. This implies that countries with a relatively larger share of global GDP per capita receive a relatively greater share of global net welfare.

According to the *vertical equity* rule, the abatement costs of a global warming policy are progressively shared across countries. This means, the country's net welfare is inversely related to the GDP. Countries with a relatively lower GDP (and thus, as a rule, with lower CO_2 emissions) receive a relatively bigger share of global net benefit (i.e. have to bear relatively less costs) than countries with a relatively higher GDP. Thus, "vertical equity expresses greater concern for the disadvantaged" (Rose 1990, p. 930). Examples for vertical equity at the interpersonal level are progressive income taxes.

The *compensation criterion* is based on the concept of Pareto-optimality. This allows a policy to be introduced only if no individual or group of individuals is worse off. The net losers have to be compensated by the net winners, so that no country suffers a negative net benefit. The concrete design of the compensation scheme requires the application of other equity criteria such as the horizontal or vertical equity principle.

Furthermore, there are two additional outcome-based equity concepts, the *basic needs approach* and the *Rawlsian Maximin rule* (see Banuri et al. 1996, p. 104). The basic needs approach grants countries the right to emit the minimum levels of greenhouse gases that safeguard the meeting of their citizen's basic needs. Basic needs can be defined, as minimum consumption levels needed to support full participation in society, depending on regional characteristics such as climate. The basic needs approach can be seen in relation to the Rawlsian approach. Taking over the Rawlsian concept of intergenerational equity on an international context, the Rawlsian Maximin rule says, "the welfare of the worst-off individual is to be maximised before all others" (Rose 1990, p. 931). This implies that this equity concept should be used to improve the position of the poorest countries. Accordingly, Rose and Stevens (1996, p. 14) operationalise this rule by redistributing any positive net benefits for the three industrialised nations in their sample to the three poorest nations. Ultimately, by giving preference to the disadvantaged group, the Rawlsian criterion is closely connected to the vertical equity principle and the egalitarian rule.

9.3.2 Allocation Based Equity Criteria

The *ability-to-pay criterion*, or more generally spoken, the principle of comparable burdens, is based on the claim that "allocation should affect all countries similarly or involve 'comparable burdens' or 'sharing the effort equally'" (Banuri et al. 1996, p. 105). Several more or less complex, operational rules of this criterion are discussed in the literature (they reach from 'equal monetary abatement costs' to more general measures of ability-to-pay). Rose and Stevens (1996) operationalise this criterion very simply by choosing an initial allocation of emission rights inversely to the proportions of the GDP.

The *egalitarian criterion* suggests that all human beings should be entitled to an equal share of the global atmospheric resource. The egalitarian rule can take two forms: equal contemporary entitlements and equal historical stock entitlements.

The first implies that the share of emissions to population is equalised, i.e. each human being should have equal rights in terms of per capita emissions. As emission rights (e.g. tradable permits or tax revenues) are distributed in proportion to each nation's current population, countries with per capita emissions below the average will gain an excess entitlement, whereas countries with per capita emissions above will have a deficit.

The second form takes 'cumulative historical emissions', i.e. emissions that have been cumulated over a period of time, as a reference point for per capita entitlements (Rose 1990, p. 929; Krumm 1995, p. 46; Banuri et al. 1996, p. 105). The second form is based on the principle of *historical responsibilities*. It takes into account that greenhouse gas emissions are stock pollutants and that the global stock, the atmosphere, is finite. From an equity point of view, it might be reasonable that not the current emissions determine the responsibility for paying, but rather the emissions, which have been built up by a country in the past. Actually, there is a broad consensus in the literature that industrialised countries have made excessive use of their rights to emit CO_2 in the past whereas developing countries are in credit. The principle of historical responsibilities is not restricted to egalitarianism, but can be used, for instance, to determine cut backs from current emission levels in direct proportion to historical, cumulative emissions.

The *sovereignty criterion* "represents the basic rights of national entities, often focused on territorial integrity" (Rose 1990, p. 930). In the case of climate change, the sovereignty rule implies equal percentage cuts of current or base year emissions. According to their baseline emission levels, countries have to undertake emission reduction measures on a different scale. The sovereignty rule is closely connected to the proportionality rule, which - as nations are burdened in proportion to their contribution to damage caused - is related to the polluter-pays principle. The sovereignty rule reflects in some respects the claim for keeping the 'status quo' allocation, as it is fully equivalent to an initial allocation of emission rights in proportion to current emissions.

Another 'status quo' rule suggests allocating emission rights proportionally to national GDP, i.e. nations with a larger share in global GDP will be better off than nations with lower shares (Krumm 1995, pp. 44). Those who defend the "emissions per GDP" criterion argue that economic activities are necessarily connected with the emission of pollutants. Thus, an allocation of emission rights that differs too much from the actual 'status quo' emission scheme might limit global production and could lead to significant reductions in global welfare. Obviously, this argument does not hold if an international reallocation of emission rights is enabled, as the equity rule restricts production of richer countries not in general but due to financial losses caused by side payments. Even so, as will be shown later in our empirical part of the paper, by considering full general equilibrium effects, this argument may have compelling meaning.

Following the literature, criteria, the sovereignty rule and the 'emissions per GDP' rule, are discussed rather as starting points in model analysis but not as serious concepts for international, global burden sharing. Accordingly, Bertram (1996, p. 468) states, that "there is general agreement in the literature that simply 'grandfathering' emission quotas on the basis of current emissions, or on the basis of present GDP, could not provide the basis of a workable international agreement, because it would impose heavy costs on non OECD countries, while enabling OECD countries to capture rents from the shortage of atmospheric carbon storage capacity for which they themselves have been responsible through high past emissions".

Under the *market justice regime* the emissions are allocated initially according to the 'willingness-to-pay'. In contrast to the burden sharing regimes discussed above, the initial allocation of permits under this rule would have to be based rather on auctioning than on grandfathering. Analogously to an international tax, the international authority, which is responsible for collecting the permit proceeds, can use an international equity rule to reimburse the revenues to the countries.

The *consensus criterion* "arises from the implications that the outcome of the political or diplomatic process is just" (Rose 1990, p. 930). Rose and Stevens (1996, p. 16), for example, operationalise this criterion by estimating weights for distributing the total amount of permits as a linear combination of each country's share of global population and its share of global GDP. The underlying principle claims that nations with concentrated economic and political power should have a greater deal of influence in the decision making process.

9.3.3 Mixed Systems

Apart from single-criterion proposals for a 'fair' distribution of emissions, a number of mixed systems have been presented in literature. Combining different equity rules might support the agreement of all countries to participate. Krumm (1995, p. 49) proposes a linked system of emissions per historical emissions, emissions per GDP and emissions per population; Welsch (1993) suggests a combination of population and current emission factors. Nevertheless, even if countries agree upon a combination of different criteria, the determination of the weight factors and their change over time remain a major subject for negotiation.

In practice, a mix of several equity concepts (namely the proportionality, historical responsibilities and ability-to-pay approach) was chosen in the Framework Convention on Climate Change. The following chapter will address this aspect in more detail.

Model simulations indicate that the speed of transition from the status quo situation to an allocation according to an internationally accepted equity rule, e.g. equal per capita emission rights, has significant influences on the international distribution of costs. Manne and Richels (1995), for example, analyse two alternative burden sharing rules which differ alone in the speed of transition from one equity principle to another and compare the impacts. Under the 'standard' allocation scheme, carbon rights are initially distributed among regions in proportion to their 1990 level of emissions (status quo). Gradual changes in these shares over time lead to a distribution in proportion to 1990 population levels by 2030 ('quicker' transition to egalitarian rule). Under the 'status quo' allocation, the gradual transition to equal per capita emission rights is slowed down and is realised not before 2200 ('slower' transition to egalitarian rule). Under the quicker transition scenario, the burden would fall on the more industrialised regions (OECD and former Soviet Union), whose share in global CO_2 emissions fall from 66% in 1990 to 22% in 2030, whereas the less industrialised countries (China and the rest of the world) would win. However, a slower transition leads to a preferential treatment of industrialised countries, which would still allow them to emit 60% of total CO_2 emissions in 2030.

9.4 The Burden Sharing Issue in the EU Context

The burden sharing issue of climate change protection arises on the global level as well as on the regional or national level. In March 1997, the Council of Ministers agreed on an allocation of 10% between the EU-member states.

Table 9-2 shows the national GHG reductions the Council decided on.

Table 9-2: The burden sharing proposal of the Council of Ministers (based on 1990 emissions, to be realised in 2010).

	Allocation of a 10% reduction in the EU
Austria	-25
Belgium	-10
Denmark	-25
Germany	-25
Finland	0
France	0
Greece	30
Ireland	15
Italy	-7
Luxemburg	-30
Netherlands	-10
Portugal	40
Sweden	5
Spain	17
United Kingdom	-10
EU-15	-10

As the Council announced, this allocation takes into account a wide range of criteria such as cost-effectiveness, differences in starting points, economic development, economic structures or resource bases (EC 1997). The proposal considers expectations about growth and technical progress (in both energy efficiency and productivity). Analysing the potential impacts of this proposal is therefore difficult, as the model framework used might be based on other assumptions concerning efficiency, growth, and so on.

To enable a comparison of the impacts of alternative equity rules on the EU-level, the equity rules analysed are kept as simple as possible. Three allocation-based criteria are evaluated with the GEM-E3 model: sovereignty, ability-to-pay and the egalitarian principle.

Supposing that a given EU-wide emission target (e.g. 90% of the EU-wide CO_2 emissions, normalised to 100) should be realised by an EU-wide permit scheme, then the figures in the columns of the following Table 9-3 indicate how many permits should be allocated to a country under a particular equity rule. The particular operational criteria used are given in the table as well.

Table 9-3: Allocation-based burden sharing in the EU.

	Sovereignty (uniform reduction rate)	Consensus (Proposal of Council of Ministers)	Egalitarianism (equal per capita emissions)	Ability-to-pay (inverse GDP per capita)
Austria	2.2	1.8	2.2	5.7
Belgium	3.9	3.9	2.9	3.3
Germany	23.9	19.7	17.8	0.1
Denmark	2.3	1.9	1.5	9.9
Finland	1.2	1.3	1.4	12.2
France	14.9	16.4	16.1	0.1
Greece	1.4	2.0	2.9	10.1
Ireland	0.8	1.0	1.0	39.1
Italy	13.6	13.9	16.7	0.1
Netherlands	3.1	3.0	4.2	1.5
Portugal	0.6	0.9	2.9	13.8
Spain	6.8	8.7	11.2	0.5
Sweden	2.9	3.4	2.4	3.6
United Kingdom	22.4	22.1	16.6	0.1
EU-14	100.0	100.0	100.0	100.0

Source: Eurostat (1994), EC (1997), own calculations.

As the table records, the burden-sharing proposal of the Council of Ministers fits in between the sovereignty and the egalitarian principle with a tendency to sovereignty. The following analysis, therefore, refrains from this particular consensus rule and emphasises more principal rules like sovereignty, egalitarianism and ability-to-pay.

Table 9-3 depicts the equity rule based allocations for a 10% reduction of EU-wide CO_2 emissions in terms of the national CO_2 emissions in 1990. The corresponding figures of the Council of Ministers proposal have been given in Table 9-2 already.

Table 9-4: Net permit allocation (difference to 1990 emissions in %).

	Sovereignty (uniform reduction rate)	Egalitarianism (equal per capita emissions)	Ability-to-pay (invers GDP per capita)
Austria	-10.0	-8.1	138.8
Belgium	-10.0	-33.5	-23.9
Germany	-10.0	-33.0	-99.7
Denmark	-10.0	-41.4	287.6
Finland	-10.0	5.4	793.9
France	-10.0	-2.6	-99.3
Greece	-10.0	84.6	542.3
Ireland	-10.0	15.9	4275.7
Italy	-10.0	10.7	-99.1
Netherlands	-10.0	24.8	-56.2
Portugal	-10.0	345.8	2019.1
Spain	-10.0	48.9	-93.5
Sweden	-10.0	-25.1	9.3
United Kingdom	-10.0	-33.4	-99.5
EU-14	-10.0	-10.0	-10.0

Source: Eurostat (1994), EC (1997), own calculations.

As indicated by Table 9-4, the amount of permits received differs considerably with respect to the equity rule chosen. Under the ability-to-pay rule Germany, France, Italy, Spain and the United Kingdom would potentially suffer with respect to the initial allocation of permits. These countries would get only a small amount of permits free-of-charge. Belgium and the Netherlands receive at least a part of the emission rights they actually would need to maintain their economic activity. The winners are low-income countries, in particular Ireland and Portugal. Under the egalitarian principle, the number of countries that contribute to the actual burden of the EU-wide emission target increases. While Austria, Belgium, Germany, Denmark, France, Sweden and the United Kingdom are discriminated with respect to the 1990 emissions, the remaining EU countries obtain more rights than needed on the 1990 basis. The sovereignty criterion gives the same reduction share to each country, i.e. each nation receives 90% of the emissions in 1990.

With respect to these allocations, the potential burden of a country (or a sector) is alleviated or reinforced. If a country gets more permits than it would actually need, the sale of these permits to other countries makes it better off (all other things being equal). If a country receives much less than it would actually need to keep its economic activity, it is obligated to either buy additional permits from other countries and/or to reduce its emissions significantly. According to the equity rules used for the initial allocation of permits, the reallocation through international trade establishes an international transfer system that favours some countries and puts others at a disadvantage.

9.5   Simulation of an EU Wide Permit Scheme under Alternative Equity Rules

The emphasis of this study lies on the analysis of the potential impacts of alternative equity rules given a particular emission target. It is not intended to contribute to the political discussion of equity and distributional fairness in general. The analysis approaches the subject from a rather positive way of thinking. Hence, the issue of concern is: If one would choose a particular initial allocation, what would be the impact for the member states and for the EU as a whole?

The analysis is based on simulation results obtained with the computable general equilibrium model GEM-E3.

9.5.1   Definition of Policy Scenario and Instrument

The policy goal imposed throughout the simulations is an EU-wide reduction of 1990 CO_2 emissions by 10 per cent in 2010. This reduction will take place by implementing an EU-wide scheme of tradable permits for CO_2 emissions in 2001. The goal is reached gradually (linear) within the following 10 years. The allocation-based rule refers to the initial allocation of permits.

It is worthwhile to explicitly mention some of the features of the permit market specified in the model.

- While international burden sharing uses different equity criteria, grandfathering (sovereignty) is used on the country level throughout all simulations. Hence, if the amount of permits given to a country covers only x per cent of the emissions in 1990, all polluters receive permits to this extent only, i.e. x per cent of their emissions in 1990 (uniform rate of reduction within the country).

- If a country obtains more permits than it actually requires (i.e. the amount of permits available exceeds the 1990 emissions), all polluters receive permits free-of-charge to the extent of their needs (actual emissions). The rest (which is the total national share minus national emission) remains with the government. This amount is supplied and sold at the international permit market. The receipts are kept by the government to reduce public deficit.

- Irrespective of the initial allocation, all polluters decide on the basis of their individual costs (marginal costs) whether to abate emissions and sell permits, to emit to the extent of the permits obtained (and keep them) or to emit more and buy additional permits. According to this decision, polluters supply or demand permits at the international market.

- The specification considers opportunity costs of holding permits, i.e. polluters take into account that even those permits that they have received free-of-charge by the initial allocation are costly as they could be sold on the market to other polluters if abatement measures are undertaken. Hence, the decision in production and consumption does not take into account the rents obtained due to the free-of-charge allocated permits. The rents a sector receives are passed on to demand by reducing the output price appropriately. The permit transactions of households are covered by the energy sectors, i.e. they are modelled similar to value added taxes with lump sum refunding.

- In any case, free trade of permits between sectors and countries guarantees a cost-minimising implementation of the EU-wide reduction target. The efficiency of the instrument (equalised marginal costs) is (theoretically) not affected by the burden sharing issue.

9.5.2 Simulation Results of Alternative Allocation Rules

The presentation of results starts with a discussion of the EU-wide effects of the three allocation rules. All simulation results presented hereafter refer to the year 2010, i.e. to the economic and environmental situation when the emission target is fully implemented.

This target is uniform in all cases: 10% reduction of CO_2 in 2010 based on 1990 emissions. The goal is reached by reducing the 10% compared to 1990 plus the emission growth that is linked to the economic growth within the period 1990 to 2010. For the EU an emission growth rate of 8% is assumed.[32] Hence, the actual reduction effort that has to be undertaken is 18%.

Table 9-5 shows the economic impacts of the three allocation principles. In terms of the EU-wide economic welfare, the sovereignty criterion is preferable. Expressed in per cent of the GDP, a positive welfare effect of 0.32% is obtained. The egalitarian rule gives a smaller but still positive effect of 0.17%. The ability-to-pay allocation reduces EU economic welfare. A loss of 0.50 (as percentage of the GDP) is indicated. Including the welfare effect that is induced by higher environmental quality improves the overall welfare effect, but the signs do not change; the welfare effect of the ability-to-pay allocation remains negative.

---

[32] Pre-Kyoto Study (Capros et al. 1997b, EIS 1997).

Table 9-5: Macro-economic impacts of alternative burden sharing allocations for EU-14.

	Sovereignty	Egalitarianism	Ability-to-pay
Gross Domestic Product	-0.80%	-0.73%	-0.42%
Employment*	-787	-683	85
Private Ivestment	-0.18%	-0.33%	-1.01%
Private Consumption	-0.10%	-0.61%	-2.64%
Exports in volume	-5.73%	-4.37%	0.16%
Imports in volume	-0.72%	-1.55%	-4.54%
Intra trade in the EU	-5.74%	-4.38%	0.15%
GDP deflator in factor prices	4.14%	2.30%	-5.28%
Marginal abatement cost*	230.5	213.9	165.1
Economic welfare***	0.32%	0.17%	-0.50%
Economic and environmental welfare***	0.49%	0.33%	-0.34%

* in thousand employed persons, ** in ECU'85/tC (ton of carbon)

*** as percent of GDP

The impacts on welfare can be explained by changes in the components of utility, i.e. consumption and leisure. While the allocations under sovereignty and egalitarianism lead to a decrease in both employment (which is equivalent to an increase in leisure) and consumption, more work has to be supplied for less consumption in the ability-to-pay allocation. Hence, utility is decreasing. The negative impact on the GDP is highest under sovereignty and lowest under ability-to-pay. The variation in GDP is linked mainly to changes in foreign trade. The GDP deflator in the EU increases by 4.1% in the sovereignty rule as all countries are affected in the same way. Due to the price increase, extra-EU exports are decreasing by more than 5.7%. In contrast, the GDP deflator decreases in the case of the ability-to-pay allocation. As it will be shown below, this drop in prices is induced by those countries which are only partly compensated by the allocation rule, i.e. which are characterised by a high GDP per capita. The different development in prices explains why the extra-EU exports remain on a higher level in the ability-to-pay case than in the sovereignty allocation. Nevertheless, the price increase under the sovereignty rule realises a reduction in EU-extra imports as the output effect dominates the substitution effect. This holds for the egalitarian rule as well.

The permit prices (marginal abatement costs) obtained at the end of the reduction period vary considerably between allocation rules. The permit price is highest under the sovereignty allocation (230 ECU/tC or 63 ECU/tCO$_2$)[33] and lowest under ability-to-pay (165 ECU/tC or 45 ECU/tCO$_2$). The outcome under the egalitarian regime is almost 214 ECU/tC (or 59 ECU/tCO$_2$). This ranking is striking, as the welfare effects are inversely related to the permit prices, i.e. a higher permit price leads to a better outcome in terms of welfare. To explain this effect, a country-specific analysis has to be undertaken.

Table 9-6 locates the sources of the EU-wide negative welfare impact in the ability-to-pay allocation: negatively affected are Germany, France, Italy, the Netherlands, Spain and the United Kingdom. The egalitarian allocation reveals negative economic welfare effects for Denmark and the United Kingdom, while applying the sovereignty rule generates positive welfare effects for all countries, even though the permit price is highest in the latter one. Receiving more permits under a particular rule is not equivalent to a more positive or less negative impact on welfare. Austria, Finland, Ireland and Sweden would prefer the sovereignty allocation, even though they would receive less permits in the initial allocation than under ability-to-pay. The reason for this effect will be made more obvious later on.

The EU-wide emission reduction in 2010 corresponds to the policy target by definition, as the endogenous permit prices match the exogenous total supply of permits (90% of 1990 emissions) and the endogenous demand of permits (i.e. the actual emissions). The actual reduction effort including both the targets related to 1990 plus the growth of emissions induced by economic growth, lies for all countries (and allocation rules) within the range of 13% to 23% (based on the emissions of 1990). The net contribution of countries to the common goal is usually much lower, depending on the growth that is assumed to take place in the different countries if no policy would take place.[34] The effects of the policy on emissions of other pollutants are not shown in the table, as the reductions are nearly stable throughout all three cases: EU-wide the actual reductions account for nitrooxide (NO$_x$) around 14%, for sulfur dioxide (SO$_2$) around 25%, for volatile organic compounds (VOC) around 9% and for particulates (PM) around 28% of the emissions in 1990.

---

[33] tC: ton of carbon, tCO$_2$: ton of carbon dioxide.

[34] As mentioned before, the growth assumptions follow the estimates given by the Pre-Kyoto Study of Capros et al. (1997b).

Table 9-6: Welfare effect and emission reduction in the EU-member states.

	Economic welfare*			CO₂ emissions**		
	Sovereignty	Egalitarianism	Ability-to-pay	Sovereignty	Egalitarianism	Ability-to-pay
Austria	0.62%	0.62%	0.56%	-15.60%	-14.79%	-12.27%
Belgium	0.65%	0.26%	0.39%	-3.71%	-3.88%	-0.91%
Germany	0.36%	0.08%	-0.56%	-25.73%	-25.92%	-25.58%
Denmark	0.27%	-0.03%	0.34%	-6.66%	-7.74%	-3.42%
Finland	0.18%	0.19%	0.16%	16.81%	17.63%	20.26%
France	0.40%	0.40%	-0.36%	-9.90%	-9.16%	-9.34%
Greece	0.10%	0.13%	0.25%	16.83%	17.57%	19.89%
Ireland	1.29%	1.24%	1.00%	4.61%	5.45%	8.03%
Italy	0.26%	0.27%	-0.58%	1.36%	2.07%	1.37%
Netherlands	0.18%	0.22%	-0.10%	9.38%	10.09%	11.00%
Portugal	0.06%	0.10%	0.22%	38.21%	39.01%	41.34%
Spain	0.38%	0.37%	-0.72%	18.99%	19.76%	18.63%
Sweden	0.47%	0.31%	0.43%	14.15%	14.39%	17.18%
United Kingdom	0.17%	-0.29%	-1.28%	-17.23%	-18.39%	-20.00%
EU-14	0.32%	0.17%	-0.50%	-10.00%	-10.00%	-10.00%

* as percent of GDP
** observed in 2010 based on 1990

Looking at the country-specific development of CO_2 emissions and the underlying emission reductions gives no clear answer to the welfare changes obtained under a particular rule. The reduction in Germany, for instance, is almost inelastic to the allocation rule, but welfare is highest under sovereignty and lowest (even negative) under ability-to-pay. While Austria or Belgium reduce more under sovereignty than under ability-to-pay, they realise a lower welfare impact under the former. For Greece and Portugal this relation is the other way round. Hence, there is no clear relation between the burden a country obtains and the actual reduction effort undertaken.

The weakness of this relation is driven by the cost-minimising behaviour of firms and households, as the principal decision of reducing emissions or keeping (or buying) permits is not altered by the amount of free-of-charge permits obtained (opportunity costs, see above). But receiving fewer permits free-of-charge reduces the ability to use these rents for output price reductions. In this case, prices remain on high levels. Hence, the less permits a country receives initially, the more the distortionary impacts of the permit scheme resemble those of emission taxes. For countries that are affected considerably (i.e. those who receive little), an international permit scheme might be even worse than a national emission tax, as the refunds are going abroad in the former, while they are kept and spent within the country in the latter.

This mechanism explains the situation indicated by the simulation results. Germany, France, Italy, Spain and the United Kingdom receive almost no permits free-of-charge in the ability-to-pay allocation. The Netherlands gets less than 50% of the 1990 emissions. All six countries show a welfare loss, as no compensation due to free-of-charge permits is available. The increase in output prices leads to a sharp decrease in domestic demand. As Table 9-7 indicates, private consumption falls in particular in those countries, which lose in terms of welfare (from 0.88% in the Netherlands to 5.37% in the United Kingdom). The same holds for the decrease in investment, which ranges from 0.44% in the Netherlands to 1.82% in the United Kingdom.

Table 9-7: Impacts on private consumption and private investment.

| | Consumption | | | Priv. Investment | | |
	Sovereignty	Egalitarianism	Ability-to-pay	Sovereignty	Egalitarianism	Ability-to-pay
Austria	0.64%	0.67%	0.65%	-0.30%	-0.19%	0.06%
Belgium	0.37%	-0.55%	-0.08%	-0.23%	-0.50%	-0.15%
Germany	-0.13%	-1.09%	-3.12%	-0.25%	-0.56%	-1.18%
Denmark	-0.34%	-1.63%	0.16%	-0.21%	-0.57%	0.09%
Finland	-0.24%	-0.19%	-0.11%	-0.21%	-0.17%	-0.08%
France	0.03%	0.14%	-2.46%	-0.41%	-0.34%	-1.00%
Greece	-0.24%	-0.16%	0.18%	0.18%	0.21%	0.30%
Ireland	1.91%	1.89%	1.56%	-0.13%	-0.04%	0.14%
Italy	-0.26%	-0.19%	-2.11%	-0.16%	-0.11%	-0.90%
Netherlands	-0.23%	-0.06%	-0.88%	-0.18%	-0.14%	-0.44%
Portugal	-0.68%	-0.56%	-0.18%	-0.84%	-0.74%	-0.40%
Spain	0.05%	0.05%	-3.46%	-0.12%	-0.09%	-1.68%
Sweden	0.34%	-0.24%	0.58%	-0.28%	-0.37%	0.09%
United Kingdom	-0.33%	-1.95%	-5.37%	0.25%	-0.40%	-1.82%
EU-14	-0.10%	-0.61%	-2.64%	-0.18%	-0.33%	-1.01%

Due to the drop in demand, domestic production for domestically produced goods and imports fall. Hence, there are two opposite effects on prices in the disadvantaged countries: The prices increase due to the purchase of permits but decrease due to the loss in demand. As the deflator of the GDP depicted in Table 9-8 indicates, the latter overcompensates the former effect in the welfare losing countries.

Table 9-8: Development of the GDP-deflator.

	GDP-deflator		
	Sovereignty	Egalitarianism	Ability-to-pay
Austria	4.68%	4.11%	1.76%
Belgium	5.11%	2.28%	0.11%
Germany	3.94%	0.86%	-6.24%
Denmark	4.24%	1.10%	1.56%
Finland	3.49%	2.82%	0.60%
France	4.74%	4.29%	-5.45%
Greece	5.28%	4.68%	2.53%
Ireland	6.76%	5.84%	2.46%
Italy	3.97%	3.51%	-5.22%
Netherlands	2.01%	1.58%	-3.08%
Portugal	1.86%	1.45%	-0.50%
Spain	5.51%	4.72%	-8.30%
Sweden	5.00%	3.12%	2.00%
United Kingdom	3.51%	-0.52%	-9.30%
EU-14	4.14%	2.30%	-5.28%

The drop in demand in the rich countries under the ability-to-pay rule explains the differences in permit prices. Compared to the other allocation rules, production and domestic demand decrease in these countries due to the higher burden they have to carry. This loss in economic activity leads to a significant reduction of CO_2-emissions. The output effect in the rich (and big) countries reduces the emission reduction effort to be reached (directly) by the endogenous permit price. Hence, even though the emission target is the same for all allocation rules, the permit price can be lower in the ability-to-pay rule.

The price depression observed under the ability-to-pay allocation in the rich (and big) countries boosts exports (see Table 9-9), which alleviates the negative effect on the GDP in the rich countries to some extent. In comparison to the trade effects of the sovereignty rule (drop of intra-EU trade by 5.74%, see Table 9-5), the differences in national price levels in the ability-to-pay allocation stimulate intra-EU trade (small increase by 0.15%).

Nevertheless, the welfare impact is less positive even in some of those countries that are favoured by the allocation rule. The reason can be found in two effects. First, the interdependence of the EU economies allows smaller economies not to make full use of the advantages they get through the ability-to-pay allocation: The negative impact on the economic performance of the big economies leads to a drop of export demand in the smaller economies, which in turn lowers the expected positive impact on welfare in the latter ones. But these spillovers can not explain the whole story, as the current account in the smaller economies is still higher under ability-to-pay than under sovereignty. Hence, a considerable part of the negative welfare effect for the supposed winners of the ability-to-pay allocation traces back to the way how the surplus of permits (i.e. those permits that are not passed on to the polluting firms but remain to the government) is used. Selling those permits on the international market and use the receipts to reduce public deficit is one way, but it has no direct impact on demand. Hence, production and income are not affected at all, which rules out positive effects on welfare. Other, more demand stimulating recycling strategies of the surplus (e.g. a lump-sum transfer to households) might be more promising if welfare losses are to be minimised. The analysis of alternative recycling strategies is a topic of future research.

Table 9-9: Impacts on foreign trade.

	Exports			Imports		
	Sovereignty	Egalitarianism	Ability-to-pay	Sovereignty	Egalitarianism	Ability-to-pay
Austria	-4.53%	-4.21%	-3.11%	-1.97%	-1.75%	-1.12%
Belgium	-3.70%	-3.04%	-2.39%	-2.23%	-2.56%	-1.72%
Germany	-4.38%	-2.43%	2.18%	-1.84%	-2.40%	-3.22%
Denmark	-3.86%	-2.49%	-2.93%	-2.76%	-3.23%	-1.61%
Finland	-3.20%	-2.93%	-2.07%	-1.66%	-1.46%	-0.92%
France	-5.22%	-5.09%	1.95%	-2.23%	-1.96%	-3.14%
Greece	-5.29%	-5.04%	-4.40%	-1.99%	-1.79%	-1.02%
Ireland	-4.71%	-4.23%	-2.77%	-1.73%	-1.38%	-0.58%
Italy	-4.30%	-4.14%	1.64%	-2.77%	-2.53%	-3.75%
Netherlands	-1.59%	-1.54%	-0.37%	-1.24%	-1.04%	-1.05%
Portugal	-1.46%	-1.49%	-1.09%	-1.83%	-1.67%	-0.74%
Spain	-5.90%	-5.53%	4.07%	-2.70%	-2.53%	-5.03%
Sweden	-4.86%	-3.77%	-3.39%	-2.57%	-2.53%	-1.24%
United Kingdom	-3.14%	-0.73%	4.70%	-1.24%	-2.23%	-4.01%
EU-14	-5.73%	-4.37%	0.16%	-0.72%	-1.55%	-4.54%

9.6 Conclusion

Allocating the burden between parties is at the focus of interest in the international negotiation process on climate protection. As the review of equity rules exhibits, the range of principles and preferences that can be applied is wide. With respect to the self-interest of nations it is, therefore, no surprise that international agreements are difficult to attain. For a comparable homogenous group of countries like the EU, the choice of the operational rule is less decisive. Nevertheless, the simulations undertaken with an EU-wide tradable permit scheme indicate that, even within the EU, the burden sharing issue matters. The ability-to-pay rule, which favours the poorer and puts the richer countries at a disadvantage, implies higher overall welfare costs for the EU than the sovereignty rule, where permits are grandfathered with respect to a uniform reduction rate. Certainly, this effect is linked to the weights given to the country-specific welfare. For the above computations no inequality aversion is applied, i.e. weights are uniform. On the other hand, the analysis of national impacts gives two insights that are not biased by this kind of evaluation.

First, if big countries (i.e. countries that are powerful in terms of economic activity) are affected considerably, the interconnection of countries through bilateral trade might make the underlying burden sharing rule less attractive even for those countries, which are favoured by a particular equity rule in terms of the initial allocation. And second, the recycling of the surplus of permits emerging for some countries in the ability-to-pay allocation has crucial impacts on consumer welfare. Using the revenues from selling this surplus on the international market to reduce public deficit leaves the rest of the economy more or less untouched, i.e. no positive signals for welfare can be found. This is the case for Austria, Finland, Ireland and Sweden, as they would prefer an allocation according to the sovereignty rule, even though they would receive less permits in the initial allocation than under ability-to-pay. Applying alternative recycling strategies may alter this result. The analysis of this aspect is a topic of future research.

With respect to economic and environmental welfare, the sovereignty rule seems to be the most acceptable for an implementation of an EU-wide permit system. All countries show a positive welfare effect and the overall EU benefit is greater than under egalitarianism and ability-to-pay.

Another, more technical insight of the analysis is linked to the differences in permit prices observed. As the initial allocation of permits might hit domestic demand (and therefore production) considerably, emissions drop due to a reduction in output. The interaction of prices and volumes generates a much lower permit price under the ability-to-pay allocation than under sovereignty or egalitarianism. Hence, the evaluation of full general equilibrium effects seems to be crucial for a consistent assessment of alternative equity rules.

10 Sensitivity Analysis

Three sets of sensitivity tests are presented in this chapter. The first two analyse the sensitivity of the results to the specific modelling assumptions adopted in the GEM-E3 model on two issues: the labour market behaviour and the reaction of the rest of the world. The last sensitivity explores the impact of exempting some or all industrial sectors and of derogating action in cohesion countries for later.

10.1 Labour Market Rigidities and the Double Dividend Issue

10.1.1 Policy Options and Scenario Definition

As an application of the GEM-E3 model we have estimated the welfare gain of alternative CO_2-reduction policies. Each policy is linked to a reduction target that is achieved by a CO_2 tax with a rate just high enough to achieve the given reduction goal. The revenue from this tax is used to reduce the employers' contribution to social security. This is the so-called "double dividend" analysis. The carbon tax should affect the substitution of energy for other inputs and reduce global warming (first dividend). This substitution effect could have already a positive impact on the demand for labour. However, the recycling of the tax money to social security as a partial compensation for employers' contribution should increase the demand for labour. The hope, the advocates of the double dividend have in mind is that the substitution effect towards more labour outweighs the output effect in terms of lower growth. Some authors describe this positive effect on the labour market as the second dividend, others define the second dividend in terms of economic welfare. Our analysis deals with both definitions as we are in general not convinced that the double dividend criterion is sufficient to be 'the one and only' measure that should be used to accept or neglect a policy. From a policy maker's view one might accept a loss in economic welfare if the policy measure supports more employment significantly.

In our empirical analysis we focus on the double dividend issue under two institutional settings. The settings are related to the degree of environmental policy co-ordination. In the first setting Germany decides to be a forerunner in combating climate change by reducing the emissions (of the base year) by 10% within ten years. The other EU countries do not apply particular measures for reduction. In the second setting the EU member states decide to combat climate change in a co-ordinated way. The target of 10% emission reduction is set at the EU level and all countries try to reach this goal together. Hence, in this setting, the marginal costs are equalised not only across sectors but across countries as well.

One could think about other schemes of co-ordination, e.g. every country reduces 10% of its base year emissions or a "full" co-ordination policy where the winners compensate the losers via side payments. These schemes are not discussed here. We rather view the climate topic from a German policy maker's perspective. Hence, the issue of concern is: "What can Germany gain from an EU-wide co-ordinated environmental tax reform?"

It is evident that the overall EU-wide emission reductions of the two scenarios specified above are different. Hence, we do not analyse solely an efficiency gain due to a better allocation from co-ordination.[35] The policies differ considerably with respect to the environmental improvement obtained and the economic burden on the economy. Comparing the two policies implicitly assumes that the German policy maker is indifferent to the actual environmental benefit achieved by the policy. One could think about such a behaviour if one finds that the policy maker reacts mainly to international requests given for example by the ongoing negotiations of the Climate Convention than to eventually occurring national environmental damages (that might be very low). Advertising a reduction of 10% could offset the international pressure on whatever level (national or EU-wide) this reduction will be. Hence, what the policy maker mind is then the national economic impact of the policy implemented and not the environmental benefit achieved. To this concern, the comparison undertaken is more a question of political opportunity than of economic efficiency.

For both policy options the emission reductions are obtained by an (endogenously computed) tax on CO_2 emissions. This tax is levied on producers and consumers. The tax reform is supposed to take place in a revenue neutral way, i.e. the receipts are kept in the country that imposes the tax and are used to reduce the employers' contribution to social security. The revenue neutrality is achieved by keeping the share of public deficit on GDP constant. As the tax and the recycling are endogenous, the problem of tax erosion induced by the decrease in emissions is implicitly solved. Hence, the reduction of wage costs through recycling of tax receipts is limited by the public deficit constraint.

---

[35] This was done e.g. by Conrad/Schmidt (1996a or 1996b).

The time horizon of target realisation is set to ten years. The reduction within these ten years is assumed to take place linearly, i.e. in each year the level of emissions (in terms of the base year) is reduced by 1%.

As the 10% target is expressed in terms of emissions of the base year, economic growth will increase the actual reduction effort. To take this aspect into account, we consider the expected emission growth in the eleven countries. The rates assumed for this study are based on an estimate of the European Commission. It is considered to be a 'conventional wisdom' scenario[36] and covers both economic growth and growth in energy efficiency. The country specific emission growth rates are depicted in the last column of Table 10-1b.

One way to consider the growth aspect in the simulations could be to calibrate energy efficiency, emission coefficients and economic growth in such a way that the model generates ('ex-post') the expected economic and environmental development. This procedure turned out to be very difficult in the complex world of our multi-country and multi-sector model.

Another way is to consider the growth aspect in the reduction plan of the policy. Then, the targets imposed include the 10% reduction of the base year emissions plus the expected emission growth.[37] Former sensitivity analyses have shown, that it is sufficient to limit the growth consideration to the policy targets if the structural changes due to the expected growth are reasonable and if one looks at the results in relative terms only. In the tables below, the policy impacts of the counterfactual equilibrium (policy simulation) are presented as percentage change of the reference equilibrium (reference simulation). What is shown then is the pure policy effect; assumptions that are kept unchanged in both scenarios are ruled out.

Furthermore, we limit the presentation of results to the impacts that are obtained at the end of the implementation horizon. In the tables below yearly values of the tenth year are depicted by a '10' in the period line while figures that concern the entire time horizon are marked by a '1-10' in the period line.[38]

10.1.2 Unilateral Action versus Co-ordination

We start with the unilateral case where Germany decides to reduce its emissions solely. The last column of Table 10-1b shows the underlying CO_2 emission growth rates assumed. The column eleven depicts an actual (net) emission

---

[36] I.e. it is not very optimistic but also not very pessimistic and includes some measures that will probably take place in the countries without imposing a 'major' policy instrument (like an emission tax).

[37] As we had no better data, we assumed that all sectors grow at the same rate.

[38] Further details on the periodical values can be obtained from the authors on request.

reduction of -7%.[39] Together with a negative expected growth rate for Germany of -3%, the 10% goal is reached. The increase of emissions in the other countries indicates the EU internal carbon leakage. The emission growth in the EU is lowered (almost 2%) but the absolute increase remains positive.

As Table 10-1b shows, the unilateral 10% reduction of Germany requires an emission tax of 8.24 ECU per tonne of CO_2 at the end of the reduction plan. The welfare effect measured in equivalent variation and expressed in percent of GDP is positive (0.02%). As both environment and welfare are improved, the so-called double dividend is obtained[40]. The welfare gain is achieved due to the increase in consumption (0.13%) which outweighs the 'negative' effect linked to the increase in employment (0.26%) (more employment means less leisure as population and therefore total time endowment is fixed). As employment and the real wage rate (0.47%) go up, there is more (labour) income for consumption. Other incomes (i.e. non labour income)[41] decrease which reinforces the effect on employment (increase in labour supply). Since gross domestic production falls by -0.20%, labour productivity declines. The investment of firms falls by -0.18%. Therefore, the positive effect in GDP is linked solely to the increase in consumption and the changes in the current account. Imports decrease (-0.24%) more than exports (-0.13%) which lead to an 'improvement' of the current account. The other countries gain or lose slightly according to the trade relations. The interested reader is referred to the figures in Table 10-1a and Table 10-1b.

---

[39] Throughout the discussion of results the sign of the numbers in the tables are kept whatever formulation is used. For example we will not change the sign in the following expressions: 'the emission growth rate is -3%' and 'the emissions fall by -3%'.

[40] The notation that is used by Goulder (1995a) distinguishes three double dividend propositions: The weak form claims for cost savings by using revenues from an environmental tax to reduce marginal rates of an existing distortionary tax compared to a situation where the revenues are transferred to households in a lump-sum fashion. The intermediate form claims that one can find a distortionary tax that would enable a (revenue neutral) substitution to an environmental tax so as to come to zero or negative costs. The strong form of the hypothesis says that there is a double dividend for typical or representative distortionary taxes, i.e. the revenue neutral substitution of the environmental tax involves zero or negative gross cost.

[41] The changes in other income are not depicted in the tables.

Table 10-1: The impact of a unilateral environmental tax reform in Germany (DE-nc) (percent changes from baseline).

a)

	EV in % of GDP	GDP (%)	Production (%)	Priv. Consumption (%)	Investment (%)	Exports (%)	Imports (%)
Period	1-10	10	10	10	10	10	10
Belgium	-0.01	0.00	-0.01	-0.01	-0.01	-0.02	-0.01
Germany	0.02	0.09	-0.20	0.13	-0.18	-0.13	-0.24
Denmark	0.01	0.00	0.00	0.01	0.00	0.00	0.01
France	-0.01	0.00	0.00	-0.01	0.00	-0.02	-0.02
Greece	-0.01	0.00	0.00	0.00	0.00	0.00	-0.01
Ireland	0.00	0.00	0.00	0.00	0.00	-0.01	0.00
Italy	-0.01	0.00	0.00	-0.01	0.00	-0.01	-0.02
Netherlands	-0.04	0.00	-0.01	-0.02	0.00	-0.02	-0.03
Portugal	0.00	0.00	0.00	0.00	0.00	-0.01	-0.01
Spain	0.00	0.00	0.00	0.00	0.00	-0.01	-0.01
United Kingdom	-0.01	0.00	0.00	-0.01	-0.01	-0.01	-0.02
EU-11	0.00	0.02	-0.05	0.03	-0.07	-0.03	-0.02

b)

	Employment (%)	After tax real wage rate (%)	Non-labour income (%)	CO_2-tax (ECU/ton CO_2)	CO_2 reduction1) (in % of base)	CO_2- emissions1) (in % of base)	CO_2- projection2) (%)
Period	10	10	10	10	1-10	1-10	1-10
Belgium	0.00	-0.01	-0.01	0.00	0.28	2.28	2.00
Germany	0.26	0.47	-0.31	8.24	-7.00	-10.00	-3.00
Denmark	0.00	0.01	0.00	0.00	0.36	17.36	17.00
France	0.00	-0.01	0.00	0.00	0.26	7.26	7.00
Greece	0.00	0.00	0.00	0.00	0.08	13.08	13.00
Ireland	0.00	0.00	0.00	0.00	0.14	13.14	13.00
Italy	0.00	-0.01	0.00	0.00	0.06	9.06	9.00
Netherlands	0.00	-0.02	-0.02	0.00	0.14	3.14	3.00
Portugal	0.00	0.00	0.00	0.00	0.06	28.06	28.00
Spain	0.00	0.00	0.00	0.00	0.00	5.00	5.00
United Kingdom	0.00	-0.01	-0.01	0.00	0.14	4.14	4.00
EU-11	0.05	-	-	-	-1.99	1.75	3.74

1) considering underlying growth (see 2)
2) projection of business-as-usual: conventional wisdom and no emission reduction policy (i.e. rates include economic growth and efficiency improvements)

The effect on employers' rate to social security is depicted in Figure 10-1. Recycling the receipts of the emission tax in a revenue neutral way leads to a reduction of this rate by 1.67 percent points (see the bar for DE-nc). As Germany acts unilaterally there is no reduction in the other countries.

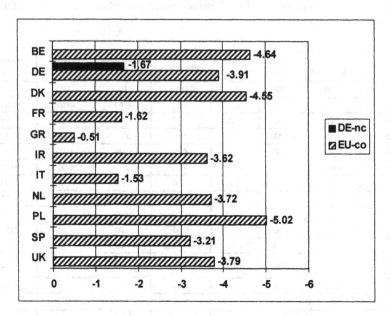

Figure 10-1: Employers' contribution to social security (change in %-points).

The simulation results of the co-ordinated policy are presented in Table 10-2a, Table 10-2b and Figure 10-1. The country specific reductions of employer's social security contribution due to the CO_2-tax receipts are depicted in Figure 10-1 (see bars of EU-co). The cuts are between 0.51 (Greece) and 5.02 (Portugal) percent points. Most countries (Belgium, Germany, Denmark, Ireland, Netherlands, Spain, and the United Kingdom) find a reduction between 3.5 and 4.5 percent points.

Real wage rates and employment increase in all countries with the exception of Greece. Greece faces a very high actual CO_2 reduction of -24.06%.

The EU-wide actual emission reduction (including the growth effect) is -13.74% while the contribution of countries to the common target differs according to their ability to adjust towards lower emissions (see Table 10-2a and Table 10-2b). The target requires a uniform tax rate of 22.01 ECU per tonne of CO_2.

The positive effect of wage cost reduction is outweighed by the decline in production (-0.78%). Hence, there is a decrease in employment (-0.06%) in Greece. Private consumption rises between 0.06% (United Kingdom) and 1.47% (Belgium). Exceptions are Greece (-0.92%, due to the decrease in real income) and Italy (-0.04%, due to a drop in other incomes that compensates the positive effect in labour income). Gross domestic production falls in all countries as the relief of labour taxes (i.e. rate of social security) is not fully compensating the increase in domestic production costs due to emission taxation. Again with the exception of Greece, GDP rises everywhere in the EU. The welfare measure EV in percent of GDP is negative for Greece (-1.71%), the United Kingdom (-0.38%), Germany (-0.22%), and Italy (-0.21%). For the other countries the EV shows up as a (positive) welfare gain.

Table 10-2: The impact of a co-ordinated environmental tax reform in the EU (EU-co) (percent changes from baseline).

a)

	EV in % of GDP	GDP (%)	Production (%)	Priv. Consumption (%)	Investment (%)	Exports (%)	Imports (%)
Period	1-10	10	10	10	10	10	10
Belgium	1.87	0.54	-0.11	1.47	0.62	-0.69	-0.24
Germany	-0.22	0.13	-0.55	0.13	-0.55	-0.37	-0.80
Denmark	0.56	0.35	-0.03	0.72	0.23	-0.44	-0.42
France	0.34	0.12	-0.27	0.34	-0.11	-0.33	-0.38
Greece	-1.71	-0.28	-0.78	-0.92	-1.32	0.21	-2.05
Ireland	1.47	0.09	-0.20	1.07	-0.08	-0.57	-0.13
Italy	-0.21	0.09	-0.30	-0.04	-0.38	-0.29	-1.01
Netherlands	0.32	0.47	-0.09	0.51	-0.01	-0.29	-0.58
Portugal	0.24	0.52	-0.29	0.67	0.05	-0.65	-0.76
Spain	0.57	0.31	-0.25	0.81	0.21	-0.41	-0.09
United Kingdom	-0.38	0.30	-0.44	0.06	-0.36	0.01	-0.81
EU-11	0.01	0.19	-0.37	0.23	-0.31	-0.22	-0.59

b)

	Employment (%)	After tax real wage rate (%)	Non-labour income (%)	CO_2-tax (ECU/ton CO_2)	CO_2 reduction1) (in % of base)	CO_2-emissions1) (in % of base)	CO_2-projection2) (%)
Period	10	10	10	10	1-10	1-10	1-10
Belgium	1.03	3.08	-0.12	22.01	-17.52	-15.52	2.00
Germany	0.58	0.85	-0.83	22.01	-15.01	-18.01	-3.00
Denmark	0.65	1.48	-0.28	22.01	-13.43	3.57	17.00
France	0.30	0.68	-0.32	22.01	-14.30	-7.30	7.00
Greece	-0.06	-1.18	-0.73	22.01	-24.06	-11.06	13.00
Ireland	0.39	1.60	0.05	22.01	-12.16	0.84	13.00
Italy	0.26	0.20	-0.58	22.01	-12.00	-3.00	9.00
Netherlands	0.97	2.02	-0.71	22.01	-14.92	-11.92	3.00
Portugal	1.17	1.97	-0.63	22.01	-13.65	14.35	28.00
Spain	0.94	1.51	-0.26	22.01	-15.25	-10.25	5.00
United Kingdom	0.63	0.48	-1.20	22.01	-10.34	-6.34	4.00
EU-11	0.54	-	-	-	-13.74	-10.00	3.74

1) considering underlying growth (see 2))
2) projection of business-as-usual: conventional wisdom and no emission reduction policy (i.e. rates include economic growth and efficiency improvements)

What are the conclusions that can be drawn at this stage of the analysis:

1. based on the parameter set used for the above two simulations, and the assumption of a perfect competitive world, we always (with the exception of Greece) find a dividend for the labour market, i.e. employment increases. This is the case irrespective of the policy imposed, i.e. co-ordinated/non co-ordinated or high/low emission reduction effort (for Germany: -7.25% in the unilateral case and -15.47% in the co-ordinated case);

2. obtaining a second dividend in terms of economic welfare is crucially linked to the design of the policy. This is shown by the German example: while the unilateral 'low' emission tax gives a positive effect on EV, the co-ordinated but higher emission tax gives an overall negative effect, i.e. the EV of Germany is negative. Amongst others there are particularly two effects of importance: First, the increasing distortion of a higher emission tax outweighs the positive effect that is linked to the decrease in labour tax and its distortionary effect. Secondly, a co-ordinated policy leads to a drop in exports as the demand in other EU-countries is now affected by own policy measures;

3. within our model framework and its parameterisation, the strong double dividend hypothesis with respect to economic welfare has to be rejected because of its lack of stability. The second dividend is linked not only to the tax level but to country specific characteristics as well. In the co-ordinated case, Denmark and Italy face approximately the same actual reduction effort under the unique tax rate, while they differ in the sign of the welfare gain: The EV is positive for Denmark whereas it is negative for Italy.

10.1.3   Labour Market Regimes

It is intuitively clear, that the imposed flexibility on the labour market plays an important role for the dividend of employment as well as for a welfare dividend.

To get an idea how sensitive impacts are due to the assumption of a perfect competitive labour market, alternative labour market regimes are tested. Imposing labour market rigidities implicitly assumes another utility function of households. For the purpose of our sensitivity analysis we assume that unemployment is completely compulsory, i.e. leisure due to unemployment is not utility increasing. For reasons of simplicity leisure demand of households is kept fixed (this is equivalent to a fixed labour supply). As an explicitly specified wage rule describes the potential development of the real wage rate, unemployment becomes a residual, i.e. supply of labour minus demand of labour.

Two imperfect labour market regimes are examined: In one variation the real wage rate is kept constant. Unions are supposed to have some bargaining power as the decrease in labour productivity linked to an environmental tax reform has no impact on the real wage rate. The nominal wage rate covers inflationary effects, i.e. it changes according to the consumer price index.

In a second analysis we assume that unions are not able to keep the level of the real wage rate but have to accept an adjustment according to the (lagged) changes in labour productivity.

Table 10-3: The impact of alternative labour market regimes (percent changes from baselines).

Labour market regime	Classical labour market						Real wage rate fix						Real wage rate according to labour productivity					
Institutional setting	Germany alone			EU (coor)			Germany alone			EU (coor)			Germany alone			EU (coor)		
Variable	EV (% of GDP)	Consumption	Employment	EV (% of GDP)	Consumption	Employment	EV (% of GDP)	Consumption	Employment	EV (% of GDP)	Consumption	Employment	EV (% of GDP)	Consumption	Employment	EV (% of GDP)	Consumption	Employment
Belgium	-0.01	-0.01	0.00	1.87	1.47	1.03	0.00	0.00	0.00	0.58	0.31	1.77	0.00	0.00	0.00	0.24	0.20	1.78
Germany	0.02	0.13	0.26	-0.22	0.13	0.58	0.08	0.04	0.51	0.08	0.05	1.11	-0.02	0.01	0.54	-0.12	-0.02	1.17
Denmark	0.01	0.01	0.00	0.56	0.72	0.65	0.02	0.01	0.02	0.53	0.34	1.44	0.03	0.02	0.02	0.38	0.28	1.50
France	-0.01	-0.01	0.00	0.34	0.34	0.30	0.00	0.00	0.00	0.20	0.10	0.63	0.00	0.00	0.00	0.06	0.05	0.65
Greece	-0.01	0.00	0.00	-1.71	-0.92	-0.06	0.01	0.00	0.01	-0.17	-0.11	0.25	0.01	0.01	0.01	-0.32	-0.15	0.24
Ireland	0.00	0.00	0.00	1.47	1.07	0.39	0.00	0.00	0.01	0.26	0.14	0.96	0.00	0.00	0.01	0.01	0.06	1.00
Italy	-0.01	-0.01	0.00	-0.21	-0.04	0.26	-0.01	-0.01	-0.01	0.00	-0.04	0.39	-0.01	-0.01	-0.01	-0.13	-0.08	0.40
Netherlands	-0.04	-0.02	0.00	0.32	0.51	0.97	-0.02	-0.01	-0.01	0.30	0.13	1.75	-0.03	-0.01	-0.01	0.05	0.05	1.76
Portugal	0.00	0.00	0.00	0.24	0.67	1.17	0.00	0.00	0.00	0.80	0.39	2.15	0.00	0.00	0.00	0.43	0.29	2.24
Spain	0.00	0.00	0.00	0.57	0.81	0.94	0.00	0.00	0.00	0.32	0.21	1.66	0.00	0.00	0.00	-0.04	0.07	1.72
United Kingdom	-0.01	-0.01	0.00	-0.38	0.06	0.63	0.00	0.00	0.00	0.07	0.04	0.95	0.00	0.00	0.00	-0.14	-0.05	0.99
EU-11	0.00	0.03	0.06	0.01	0.23	0.54	0.02	0.01	0.10	0.14	0.07	0.97	-0.01	0.00	0.11	-0.05	0.00	1.00

Table 10-3 shows the findings of this sensitivity analysis with respect to equivalent variation (EV) as percentage of GDP, consumption and employment. Note that the EV is an ordinal measure, which can be used to evaluate policy simulations within one model variation only. As the rigid labour market regimes implicitly assume a different utility evaluation, the EV can show which policy is preferred under a particular labour market regime but it is not appropriate to assess policies under different labour market regimes:[42] While there is no unemployment in a classical labour market, leisure of the unemployed is supposed to be involuntary in the rigid labour market regimes. As in the latter more employment is matched by former unemployed, there is no loss in utility because of a decrease in leisure. Hence, to compare the results across model variations one has to take the two components of utility, i.e. consumption and employment (respectively leisure) into account.

In the classical labour market regime, Germany would prefer the national policy. There is a welfare gain for the households of 0.02% of GDP, whereas the co-ordinated policy yields a welfare loss of -0.22% of GDP. If real wages are assumed to be inflexible (fixed real wage rate) German households are indifferent to the institutional setting of the policy but they would like to see one of the two implemented: the EV of both is equal and positive. For a wage rule with labour productivity adjustment, Germany is better off to choose the national policy, but German consumers would like to see none of the two policy options to be implemented.

To enable a comparison across model variations changes in consumption and employment have to be taken into account. We will exemplary discuss the impact of different labour market specifications for the unilateral policy[43]: If real wage rates are fixed, demand for labour is higher than in the flexible labour market as there is no price mechanism that adjusts demand and supply (which would lead to higher wage costs). The additional labour demand is met by the formerly unemployed. Production costs and the fall in total output are lower. At the same time private consumption rises less than in the flexible labour market case as the increase in real income is lower. Hence, households have to work more for an additional unit of consumption than in the classical labour market case. However, the distortionary effect of the pre-existing labour tax (rate of social security contributions) is higher in the case of unemployment,[44] which is another source of the positive effect in employment. In the third labour market specification the decrease in labour productivity lowers real wage rate. The effect on employment is even a bit higher than in the case with fixed real wage rate. But private consumption is almost unchanged, as the level of real income is not affected considerably; the employment effect is outweighed by the decrease in the real wage rate.

---

[42] The main reason for this inconvenience is the fact that we have to compute a separate reference run for each labour market regime. As the computations of EVs refer to different bases they are not comparable.

[43] For reasons of clarity we suppress a more detailed presentation of the figures of these simulations.

[44] This is in line with the findings of Bovenberg/de Mooij (1994b) and Bovenberg/van der Ploeg (1994).

How do the results of the co-ordinated policy differ with respect to the different assumptions for the labour market? In principle the findings are similar to those for the unilateral policy: there is more employment under a rigid labour market but real income and therefore consumption increases less. The marginal rate of consumption to employment is highest for the flexible labour market, decreases in a market with fixed wage rate and is lowest for a labour market where the real wage is adjusted according to labour productivity. Hence, in terms of households' efficiency the rigid labour market regimes lowers the positive effect of the policy.

If the goal of the policy is to increase employment without lowering the level of consumption, the co-ordinated policy is appropriate for most countries. Under the labour market regimes 'perfect competition' and 'fixed real wage rate' only Greece and Italy are affected negatively, whereas in the third regime Germany, Greece, Italy and the United Kingdom are worse off. Under all three market regimes, there is an EU-wide positive effect for labour without lowering the total level of consumption.

What are the Conclusions of the Labour Market Sensitivity Analysis?

Labour market rigidities alter the results found in a perfect competitive world. But again everything depends: Labour market rigidities will worsen the impact of the policy on employment in those model applications, which find no dividend for employment in a competitive world, as less flexibility in the labour market will increase adjustment costs. Whereas a rigid labour market improves the (positive) effect on labour in applications that find a dividend for employment in the perfect world already. This is obvious as a positive effect on employment in a competitive world is linked to the increase in the after tax real wage rate. A wage rule that keeps the real wage rate fixed enables firms to demand the formerly unemployed labour force at unchanged unit costs. Hence, production costs are lower than in the competitive case.

It is beyond the scope of this paper to discuss the appropriateness of the labour market regimes imposed and to judge about the most realistic one. Nevertheless in undertaking a medium term analysis one might expect some inflexibility in real wages. A decrease in real wages or the assumption of a clearing labour market seems at least in the short term less plausible. If one believes in the inflexibility of wages in the short and medium term, a co-ordinated policy that taxes emissions and reduces wage costs would be favourable for all countries with the exception of Greece. The other labour market regimes traces more losers, i.e. there are more countries that refuse the mitigation policy because an economic dividend is not obtained. Another aspect is the question of compensating the losers. As the figures in the tables show, there is an EU-wide benefit of co-ordination under the flexible labour market and under the regime with fixed real wage rate. If one would find a compensation mechanism that makes the losers better off (at least indifferent to the policy), one might find a strategy where all countries are willing to cooperate. This aspect is a topic of future research.

10.1.4 Exploring the Deviancy

Why are our empirical results not fully compatible with some other theoretical and empirical findings of e.g. Bovenberg/de Mooij (1994a, 1994b), Pethig (1996) or Goulder (1995b), who reject the double dividend hypothesis more or less systematically? In summary the differences are linked to the following issues:

- full general equilibrium effects versus comparative static analysis,

- agents considered,

- endogeneity of factors and number of inputs considered,

- optimality of the pre-existing tax system,

- assumed factor mobility,

- foreign trade specification and its parameterisation.

Analytical studies done e.g. by Bovenberg/de Mooij (1994a, 1994b), Bovenberg/van der Ploeg (1994) or Pethig (1996) are based on more simple static general equilibrium models. Their work focuses on an analytical comparative static evaluation of these models. As Pethig stresses, *comparative static analysis* becomes "... very messy even in simple models..." if one releases the very restrictive assumptions on preferences and technologies that enable a full characterisation of conditions under which a double dividend might be obtained. Applied general equilibrium analysis allows for more complex specifications, which cannot be solved analytically any more. Hence, it is a priori not clear, why the results obtained by comparative static analysis in simplified models should hold in a much more complex world that uses real world data and traces *full general equilibrium effects*.

Another explanation is the *consideration of agents*. In contrast to the theoretical analysis of Bovenberg and others, where only three agents are distinguished (households and firms are not separated), the model used here takes four agents into account (household, firms, government and foreign). Therefore firms keep a part of capital income for investment. Hence, a policy-induced decrease in capital income affects both, firms and households, which in turn lowers the (negative) impact on private consumption.

Furthermore, the theoretical result in Bovenberg/de Mooij (1994a), that pollution taxes reduce the incentive to supply labour, is not in contradiction to our model results because their proof is based on the *assumption of a single input* (labour) and on a constant labour productivity. In Bovenberg/Van der Ploeg (1994) three inputs are used (labour, energy, and capital), however the prices of capital and energy are determined on global competitive markets, i.e. they are exogenous. In their factor price frontier $\left(w+t_l\right)\big/_P = \phi\left(\dfrac{\overline{PK}}{P}, \dfrac{\overline{PF}+t_F}{P}\right)$, a given tax on fossil fuel

(t_F) uniquely determines the producer wage $\frac{(w+t_L)}{P}$. Hence, the energy tax is fully born by the immobile factor labour and thus amounts to an implicit labour tax. The factor price frontier in our model, derived from the unit cost function, is (in a shorter version):

$$1 = \phi\left(\frac{PL}{P}, \frac{PK}{P}, \frac{(PF+t_F)}{P}, \frac{PM}{P}\right),$$

where prices of capital and of energy are endogenous. The carbon tax is therefore not an implicit labour tax, i.e. the effect of a lower tax on wages is not fully offset by the carbon tax. This explains why in Bovenberg/Van der Ploeg an increase in the energy tax harms employment as the higher energy tax is shifted onto the immobile factor (i.e. labour) only. Hence, the higher cost of energy depresses the market wage $\frac{w}{P}$. In the GEM-E3 model it depresses both factor prices, i.e. labour and capital, as the user cost of capital is endogenous. In an open economy framework the endogenous PF absorbs also some of the tax impact (depending on the 'openness' assumed).

The GEM-E3 model includes a range of *pre-existing taxes* and transfers, e.g. direct taxation of households and firms, employers and employees contribution to social security, import duties, subsidies, value added tax and social benefits. In contrast to the assumptions of e.g. Bovenberg/de Mooij, the pre-existing tax systems in the countries considered are not optimal. Hence, one could argue that a tax reform could enable efficiency gains, which are not linked solely to a new environmental tax but to a general change in (effective) factor tax rates.[45] But this is still not in line with the findings of Bovenberg/Goulder (1996), who show that both, the analytical and the empirical analysis coincidence even if one considers pre-existing taxes: While they find analytically that the prospects of a double dividend are enhanced if "... a revenue neutral tax reform shifts the burden of taxation to the less efficient (undertaxed) factor...", there is no empirical evidence obtaining such a situation: no double dividend is obtained in their numerical analysis for a wide range of parameters.

Again these findings are crucially linked to the *factor mobility assumptions* they make. In both, the analytical and the numerical analysis capital is fully mobile. While in a comparative static analysis capital mobility is usually assumed to cover the long term effects of a policy, the meaning is different if it is imposed for an intertemporal (periodically computed) numerical model. There the assumption of full capital mobility holds for both, the short (which seems to be rather unrealistic) and the long run. It is worthwhile to stress this aspect as the mobility issue has important impacts on the excess burden of taxes. As Böhringer et al. (1996) show

---

[45] Goulder (1995b).

for their CGE-model for Germany, the marginal excess burden of the overall tax system decreases sharply if the assumption of full capital mobility is removed. As Bovenberg/Goulder (1996) concede, a highly sub-optimal pre-existing tax system (i.e. a tax system where the marginal excess burden of different taxes vary quite a lot) enhances the chance of obtaining a double dividend, if the burden of the new environmental tax falls mainly on the undertaxed factor (which is capital in our case) and relieves the burden of the overtaxed factor (labour).

The putty-clay approach used in our recursively dynamic model assumes sectoral fixed capital stocks within a single period. In such a situation the burden of the environmental tax falls partly on capital as stocks can adjust only gradually over time by depreciation and gross investment. This specification is in line with the empirical evidence that can be found at least in the short and medium term and which denies that firms can easily grab their facilities and go somewhere else if production conditions become a bit worse than they were before.[46]

Of course there are other differences in model specifications that might play a role for the results obtained. Even if the differences are of minor importance in a comparative static analysis they might not be negligible in a full equilibrium analysis based on real data. Alternative specification and differences in the underlying data are the major reason why the findings of empirical work are less unique than those of the analytical work. While the models of Goulder (1995a) and Bovenberg/Goulder (1996) trace no double dividend for the US economy, the models of Jorgenson/Wilcoxen (1994) and Ballard/Medema (1993) confirm the double dividend hypothesis for the same economy.

Finally, one should reconsider that the analysis traces economic benefits or losses only. The welfare measure chosen abstracts from any (individual) environmental benefit. As the simulation results indicate, a tax on CO_2 reduces also emissions of other (energy-related) pollutants like SO_2 or NO_x. The sum of benefits from slowing environmental degradation might outweigh occurring costs easily. Integrating economic costs and environmental benefits within one assessment framework is also a topic of our future research.

10.1.5 Conclusion

National economic concerns still hinder the implementation of effective policy measures that would enable a significant reduction of greenhouse gas emissions. Considering other economic problems like unemployment, the enforcement of Maastricht criterions and highly tensed social security systems, the global warming problem seems to be of less priority for national policy makers. Nevertheless, the international political pressure forces governments to do

---

[46] See e.g. Dean (1992), Jaffe et al. (1995).

something and show at least political will. To escape the dilemma countries choose their strategy with respect to least cost criteria; the potential environmental benefit is of less importance as long as the reductions achieved relieve the political requests of the international society.

The purpose of this study was to point out clearly the assumptions underlying the model specification and their potential impact on the empirical outcomes of the analysis. There is a wide range of factors that play an important role for the double dividend issue. Apart from the labour market specification this concerns in particular the 'openness' of the economy assumed, the number of agents considered and the factor mobility imposed. While alternative assumptions with respect to these issues might explain some of the divergence of results that can be found in the literature, a major disadvantage remains: there is (at least at the current state of economic modelling) no complete reliability either to reject or to confirm the existence of a double dividend - everything is linked to the reliability of the model specification and its parameterization.

Our simulations with a multi-country multi-sector CGE-model for the European Union indicate that a revenue neutral environmental tax reform, where CO_2 emissions are taxed and the tax receipts are used to release the employers' contribution to social security, might yield a double dividend. But the positive effect obtained on the economic side is rather small (for Germany e.g. the positive welfare could be 0.02 percent of GDP) and it depends crucially on the institutional setting of the policy chosen (unilateral versus EU-wide co-ordination) as well as on the labour market regime imposed (flexible versus rigid labour market).

In a classical labour market specification Germany obtains a double dividend if a unilateral policy is imposed. The co-ordinated policy makes Germany worse off as a small welfare loss is obtained. There are two reasons for explanation: First, the two policies (unilateral: 10% CO_2 reduction in Germany, co-ordinated: 10% CO_2 reduction EU-wide) are not neutral with respect to the environmental goal obtained and second, the burden sharing between countries in the co-ordinated policy is not considering the 'polluter pays principle.' The allocation of a country's contribution is obtained by the (co-ordinated) minimum cost solution on the EU level. Taking into account the differences in potential[47] emission growth rates (which is below average for Germany) yields a reduction for Germany that is much higher in the co-ordinated case compared to the one by unilateral action.

Nevertheless there is an overall benefit of co-ordination. Therefore, more sophisticated co-ordination schemes that include side payments could enable the acceptance of such a co-ordinated policy by all parties (countries).

The outcomes are different if one specifies the labour market alternatively. Assuming fixed real wage rates makes Germany indifferent to the two policies;

---

[47] If no policy action would have been undertaken.

there is a small welfare gain in both policies. If real wage rates are adjusted according to the decrease in labour productivity - the latter is linked to every policy that reduces labour cost - no double dividend is obtained. Hence, justification or rejection of the double dividend hypothesis depends crucially on the labour market assumptions made.

From an empirical point of view, obtaining an economic dividend in terms of welfare by undertaking an 'intelligent' environmental tax reform remains at least possible, but relying on it seems rather optimistic, as there are numerous uncertain influences that might alter the sign of the welfare effect. Obtaining a second dividend in terms of employment seems to be more robust. At any rate, even if there might be no 'no regrets' policy, the economic impact of such a reform seems - from the perspective of consumer welfare - within acceptable limits.

10.2 World Context

10.2.1 Introduction

In open-economy applied general equilibrium models the specification of foreign trade and of the behaviour in the rest of the world (RoW) is an important feature. In the literature a distinction is drawn between multi-country and single-country models[48]. While the former are mainly designed to analyse global issues, the latter takes the perspective of a single country. Multi- and single-country models differ also with regard to the modelling of trade determinants, i.e. in the way of modelling export and import behaviour. In multi-country models, or world models respectively, production and demand are specified for all countries participating in trade. All regions covered by the model are linked together by bilateral world trade matrices or trade pools. Compared with that, in single-country models the behaviour in the RoW is modelled rather roughly. Typically, a 'closure rule' for trade with the external sector is incorporated, i.e. a *crude* specification of the RoW's import-demand and export-supply functions which is usually completed by a balance-of-payments condition (Shoven and Whalley 1992, p. 81).

As recent studies indicate, the closure rule chosen in a general equilibrium model, and thus in the GEM-E3 model as well, may be of particular importance for simulation results49.

---

[48] Shoven and Whalley (1992) give an overview on recent multi-country and single-country models.

[49] Whalley and Yeung (1984) examine how results from policy simulations depend on the assumptions about international trade, using a simple numerical example. The external sector specifications vary according to the elasticity of the foreign offer curve. They

The GEM-E3 model includes 14 EU countries and the RoW covering all other industrialised regions and all developing countries. It is not a global model, as the behaviour of the RoW is kept exogenous in large parts. World production and export prices are fixed, i.e. export supply is assumed to be perfectly price elastic. This assumption reflects price-taking behaviour of the EU vis-à-vis RoW. But, as price-taking behaviour is accompanied by product differentiation due to the Armington assumption, the price level in the European Union is not completely determined by the world market (and exchange rates). Thus, an exogenous rise in foreign export prices would affect the EU-wide price level only partly.

Another important aspect in the GEM-E3 model is the modelling of interactions between macroeconomic developments in the EU and the foreign sector. Actually, the only feedback is a price elastic foreign demand for EU exports. Optionally, for the long-term analysis, an additional feedback mechanism can be introduced by a balance-of-payments constraint.

Basically, the assumption that the export prices of the RoW remain constant, independent from the amount of imports demanded by the EU, is rather restrictive. Thus, it seems reasonable to relax the small-country assumption and to assume that trade activities of the EU affect world prices.

The objective of this chapter is to clarify the relationship between the foreign sector and the EU economy in the GEM-E3 model. For reasons of simplicity, the analysis is based on the (real) standard version of the GEM-E3 model where money and asset markets are excluded. This has the advantage that any policy-induced change is fully reflected by changes in real variables, and is not absorbed by money market effects.

10.2.2 Sensitivity to Foreign Trade Specifications

In this section sensitivity analysis will be conducted with respect to alternative foreign trade specifications. Basically, three approaches will be tested:

- an additional price equation for exports from RoW to EU is introduced. Instead of fixed world prices for exports, the EU is faced with a finite price elastic export supply function;

- the foreign import demand function is changed by introducing a link between the activity level of the domestic (EU) and the foreign (RoW) economy;

---

include as extremes the large country assumption and the small, price taking country formulation in which the country has only marginal influence over its terms of trade. Calculating the equilibrium effects of a distorting capital tax Whalley and Yeung yield a substantial sensitivity of results in terms of welfare gains to the external sector specification. Whereas in case of the large country assumption the terms of trade loss offsets the gain from the removal of a distorting tax, in case of the small country assumption the domestic gain is at its highest.

- variations in the degree of substitution between goods entering the sectoral aggregate import demand functions of both EU countries and RoW are analysed.

The whole sensitivity analysis will be based on the co-ordinated double dividend scenario. As mainly short- and medium-term aspects are considered, the balance-of-payments is kept variable. Policy-induced impacts are calculated for all variations in the foreign trade sector suggested above. The sensitivity of results is analysed by comparing the results with those produced by the (unchanged) standard version of the GEM-E3 model. For reasons of clarity, the discussion of results concentrates on selected EU-14 macroeconomic and sectoral aggregates.

10.2.3 Changes in Rows Export Supply

10.2.3.1 Specification of Row's Export Supply

In this section the assumption of a perfectly price elastic export supply function of the RoW is given up. Instead, for each sector a foreign export supply function with constant own-price elasticity is introduced:

$$(1)\ PEX_{row} = \left(\frac{EX_{row}}{EX_{row,0}} \right)^{\frac{1}{\gamma}}, \quad 0 < \gamma < \infty.$$

In the following, equation (1) is introduced as an additional price equation for all sectors in the GEM-E3 standard model version. Now prices of exports from RoW are no longer fixed, but increase with the amount of RoW's exports, or, with the amount of EU-14 imports respectively. Obviously, introducing this new specification can lead to substantial changes in simulation results, in particular if the policy induced impacts on EU imports are substantial.

The new specification is tested for three alternative parameter values of γ (see Table 10-4). For reasons of simplicity, γ is not differentiated among sectors.

Table 10-4: Values of parameter γ for sensitivity analysis.

Sector	Case 0: Standard version of GEM-E3	Case 1: Halved values	Case 2: Central values	Case 3: Doubled values
1 - 18	∞	0.5	1	2

10.2.3.2 Simulation Results

The following double dividend simulations include the case of a perfectly elastic export supply function (reflecting the standard version of GEM-E3) and the case of not perfectly elastic export supply functions as specified in the previous section. Results from these cases are reported in Table 10-5 in terms of several macroeconomic aggregates.

The results indicate that, contrary to expectations, the EU-14 as a whole would gain from more flexible export prices in terms of economic welfare. The lower the own-price elasticity of foreign export supplies the higher the economic welfare. While in the standard version of the GEM-E3 model (fixed export prices) the welfare effect of the double dividend policy is around 0.23%, it rises to 0.32% in Case 3, to 0.42% in Case 2 and, finally, to 0.62% in Case 1.

Overall, gross domestic product, employment, production, private investment, private consumption, extra-EU imports and energy consumption are higher, the lower the own-price elasticity of export supply is. For example, gross domestic product drops in the standard version ($\gamma \rightarrow \infty$) by -0.04% and still in Case 3 ($\gamma = 2$) by -0.01%, but, however, rises in Case 2 ($\gamma = 1$) by 0.01% and in Case 1 ($\gamma = 0.5$) by 0.05%.

The impacts on exports are opposite to those described above. Exports run parallel to the value of the own-price elasticity of export supply. The volume of intra trade in the EU reacts in the same way. Intra-EU trade, defined as intra-EU exports, decreases the most in Case 1 and the least in the standard version.

As already mentioned, the new specification leads to a greater fall in EU-14 exports. This is due to an additional increase in EU production costs, which is caused by higher prices for RoW's exports. Producers in EU now have less possibility for cushioning the tax induced EU-wide price increase.

As Table 10-5 shows, the positive employment effects are stronger in case of flexible RoW's export prices. This can be easily understood taking into account that in particular energy-intensive sectors for which domestic prices rise considerably will be substituted by an increasing extent by imported goods from RoW or by input factors with relatively lower prices, such as labour. As cost-effective possibilities of a switch to foreign products are restricted, the switch to labour is reinforced. As labour demand rises stronger, real wage rates are pushed up to a greater extent as well. Thus, labour supply and employment increase. Rising income stimulates consumption of private households, which in turn reduces the negative impact on production.

Imports are higher compared to the standard version and take a positive percentage rate for $\gamma = 0.5$. The rise in import demand is driven by a lower decrease in production (imports of non tradable and non-competitive imports) and increasing consumption (imports entering the Armington demand function). The positive effect on consumption outweighs the loss in leisure; hence, welfare increases.

Table 10-5: Co-ordinated double dividend policy constant price elastic foreign export supply.

	Macroeconomic Aggregates for EU-14			
	Case 0: Standard version of GEM-E3	Case 1: Halved values (γ=0.5)	Case 2: Central values (γ=1)	Case 3: Doubled values (γ=2)
Gross Domestic Product	-0.04%	0.05%	0.01%	-0.01%
Employment*	780	1108	950,	868
Production	-0.57%	-0.50%	-0.54%	-0.56%
Private Investment	-0.18%	-0.01%	-0.09%	-0.13%
Private Consumption	0.21%	1.03%	0.61%	0.41%
Domestic Demand	-0.56%	-0.26%	-0.41%	-0.49%
Exports in volume	-1.02%	-2.64%	-1.81%	-1.41%
Imports in volume	-1.46%	0.04%	-0.74%	-1.11%
Intra trade in the EU	-1.20%	-2.20%	-1.71%	-1.47%
Energy consumption in volume	-6.21%	-5.80%	-6.01%	-6.12%
Consumers' price index	1.19%	6.51%	3.02%	1.92%
GDP deflator in factor prices	-0.74%	4.50%	1.05%	-0.04%
Current account as % of GDP***	0.08	0.20	0.14	0.11
Equivalent Variation of total welfare				
Economic welfare**	0.23%	0.62%	0.42%	0.32%
* in thousand employed persons				
** as percent of GDP				
*** absolute difference from baseline				

(Numbers indicate percent changes from baseline except if defined otherwise).

10.2.4   Changes in Row's Import Demand

10.2.4.1 Specification of Row's Import Demand

In the standard version of the GEM-E3 model neither production and consumption nor domestic supply in RoW are endogenous. Domestic demand for domestically produced goods, which enters RoW's import demand function, is given, too. Hence, no linkage to the economy's activity level is incorporated in RoW's import demand specification. Thus, in contrast to the EU import demand specification, import demand of RoW depends alone on relative prices (terms-of-trade).

The idea behind the specification presented below is to introduce an additional endogenous variable in the foreign import demand function, which measures the economic performance of RoW. As in the standard version of the GEM-E3 model production of RoW is fixed, RoW's exports are used as 'activity variable' entering import demand. However, as RoW's actual exports are completely determined by import demand of EU-14, RoW's import demand is no longer influenced exclusively by EU country-specific export prices, but also by the amount of imports demanded by EU-14.

The specification represents a rough attempt to provide RoW's import demand function with more flexibility and empirical evidence as economic interactions between the two regions are now taken into account. If in reality the economy of EU-14 expands and income rises, EU-imports will rise, too, because a part of additional income will be spent for additional imports. This in turn implies a rise in exports of RoW. Up to this point, the interactions are covered by the standard model version. However, in addition to this, in reality an increase in RoW's exports would result in an increase in RoW's income, and thus in an increase in RoW's import demand as well. This feedback mechanism which is ignored in the standard model version has been included in the new specification presented in the following.

The specification of XXD_{row} will be changed by relating RoW's production to RoW's exports. First of all, we assume that production in RoW XD_{row} is a function of RoW's exports EX_{row}

$$(2)\ XD_{row} = \beta \cdot (EX_{row})^{\varphi}.$$

φ may be interpreted as elasticity of RoW's production to RoW's exports which measures the degree of linkage between EU and foreign economy. It is assumed that the share of RoW's exports in RoW's production, θ, and thus the share of domestically-sold and domestically-produced goods in domestic production, $(1 - \theta)$, is fixed. With this assumption:

$$(3) \quad IMP_{row,k} = (EX_{row})^{\varphi} \cdot \alpha_k \cdot \left(\frac{PXD_{row} \cdot e_k}{PEX_k \cdot e_{row}} \right)^{\sigma^{row}} \quad \forall k, \quad k = 1, \ldots, 14,$$

where $\alpha_k = (1 - \theta) \cdot \beta \cdot \delta m_{row,k} \cdot \dfrac{\delta x_{1,row}}{\delta x_{2,row}}$ is calibrated to the observed benchmark data.

In order to specify φ equation (3) is used as a regression equation with RoW's exports as an explanatory (exogenous) and RoW's production as a dependent (endogenous) variable. The regression coefficients β and φ are estimated by the least-squares method.

Table 10-6: Sectoral values of parameter φ for sensitivity analysis.

Sector	Case 0: Standard version of GEM-E3	Case 1: Halved values	Case 2: Central values	Case 3: Doubled values
1 - 5, 13 - 18	0	0.22	0.43	0.86
6 - 8	0	0.24	0.47	0.94
9 - 11	0	0.29	0.57	1.14
12	0	0.13	0.25	0.50

10.2.4.2 Simulation Results

Simulation results of the double dividend scenario are reported in Table 10-5 for macroeconomic aggregates. Obviously, the introduction of a linkage between production and exports has just slight impacts on model results. Impacts are the greatest for Case 3 where the feedback parameter φ takes the highest values.

With respect to gross domestic product no changes can be observed. The percentage reduction rate of -0.04% stays the same in all cases. Anyhow, employment, production, private investment, private consumption, domestic demand and exports show some small changes compared to the standard case. In particular, the decreases in imports (-1.24%) and in GDP deflator in factor prices (-0.38%) are cushioned slightly in the third case. EU-wide economic welfare is not much affected by a variation of φ at all. However, economic welfare as percentage of GDP increases in Case 3 by 0.25% compared to the reference scenario. Compared to the standard version, economic welfare rises by 0.02 percentage points or by 10% respectively.

These differences in results can be explained as follows. As explained above, the double dividend policy brings about a rise in production costs and in consumer prices for domestically produced goods. As a result, domestic demand, and exports as well, are reduced. This reduction is, together with the decline of exports, responsible for a decrease in an overall production level in EU-14. Thus, the quantity of imports demanded, or the quantity of RoW's exports respectively, is reduced as well, as the substitution effect from domestic to foreign goods is not big enough to compensate the negative income effect. According to the new import demand specification, production in RoW decreases if RoW's exports go down, and thus RoW's imports, or EU-14 exports respectively, go down, too. In the end, aggregate exports of EU-14 are forced back stronger if RoW's import demand is modelled dependently on RoW's exports. According to Table 10-5, in the standard case exports fall by -1.02% only, while in Case 1 to 3 they are reduced by -1.04%, or -1.05% respectively. To summarise, in the changed model version a reduction of EU-14 imports has a negative effect on imports of RoW, assuming other things being equal. This negative effect is growing with φ, i.e. with the link between RoW's exports and imports.

However, in contrast to its impact on aggregate exports, the new specification tones down the fall of aggregate imports. While imports are reduced by -1.46% in the standard case, they show a slightly less decrease when the new specification is applied. In particular, in Case 3 imports fall by -1.24% only. This pattern is also evident for the development of sectoral imports, apart from few exceptions (e.g. sector coal and oil). The increase in imports (with growing values of φ) can easily be explained by the rise of the consumer price index. While the consumer price index changes by 1.19% in Case 0, it goes up to 1.22% in Case 1, to 1.27% in Case 2 and to 1.55% in Case 3. This increase is explained by the reduction in EU-14 exports due to the setting of φ. As prices of exported goods contain to a certain degree tax payments which have been paid by European producers, the lower the exports, the lower the share of tax burden that can be shifted indirectly to abroad. Consequently, if exports go down, European consumers themselves have to bear a greater part of tax burden. This is reflected in increased consumer prices. At the sectoral level, a greater shift in the structure of imports can be observed.

All in all, the impact of the changed import demand specification is very low. Certainly, impact will be stronger if higher values for φ are chosen. But as the EU-14 is a small economy compared to RoW, the strength of the feedback between EU imports and RoW imports should not be overestimated. While the specification must be interpreted with reservations at all, there is in particular less evidence to support it for higher values of φ.

Table 10-7: Co-ordinated double dividend policy changed import demand of row.

Macroeconomic Aggregates for EU-14				
	Case 0: Standard version of GEM- E3	Case 1: Halved values	Case 2: Central values	Case 3: Doubled values
Gross Domestic Product	-0.04%	-0.04%	-0.04%	-0.04%
Employment*	780	784	789	805
Production	-0.57%	-0.58%	-0.57%	-0.58%
Private Investment	-0.18%	-0.17%	-0.17%	-0.15%
Private Consumption	0.21%	0.21%	0.21%	0.26%
Domestic Demand	-0.56%	-0.56%	-0.55%	-0.54%
Exports in volume	-1.02%	-1.04%	-1.05%	-1.05%
Imports in volume	-1.46%	-1.46%	-1.44%	-1.24%
Intra trade in the EU	-1.20%	-1.23%	-1.27%	-1.36%
Energy consumption in volume	-6.21%	-6.24%	-6.27%	-6.31%
Consumers' price index	1.19%	1.22%	1.27%	1.55%
GDP deflator in factor prices	-0.74%	-0.71%	-0.67%	-0.38%
Current account as % of GDP***	0.08	0.08	0.08	0.08
Equivalent Variation of total welfare				
Economic welfare**	0.23%	0.23%	0.23%	0.25%
* in thousand employed persons ** as percent of GDP *** absolute difference from baseline				

(Numbers indicate percent changes from baseline except if defined otherwise).

10.2.5 Variation of Armington Elasticity Values

Armington elasticities at the lower and upper level of substitution may represent a key parameter in simulation models as, they affect substitution possibilities between imported and domestically produced goods (see de Melo and Robinson 1989, p. 57ff.). In particular, they influence the strength of terms-of-trade effects and, along with production and consumption effects, they determine the total welfare change of any policy measure (see Whalley 1985, p. 110). Thus, a critical issue in CGE modelling is the choice of elasticity values. Whereas in the GEM-E3 model share parameters are calibrated to the base year's observed data set, the values of sector- and country specific substitution elasticities have to be specified from the outside of the model. This is due to an under-identification problem of the calibration procedure as the benchmark data set alone is not enough to determine all parameter values (Fehr et al. 1995, p. 151).

10.2.5.1 Variation of Armington Elasticities: Row

10.2.5.1.1   Specifications of Armington Elasticities

Table 10-8 considers four variations of upper-level substitution elasticities, which are used for subsequent sensitivity analyses. The first column contains the values of the standard version of the GEM-E3 model (Case 0), around which sensitivity analysis is performed. In Case 1, all sectoral elasticity values are halved from those used in the standard model. In Case 2 values are doubled. The fourth and the fifth column depict 'best guess' estimates (Case 3) as well as econometric estimates (Case 4), both taken from the Shiells et al. study (1986). As this study is based on the three-digit ISIC classification, the values have been aggregated according to the GEM-E3 18-sector scheme using 1988 RoW's import shares as weights. Ultimately, U.S. literature-based estimates are taken as crude proxy for the RoW's behaviour. Unfortunately, the database, provided by Shiells et al. is not sufficient to calculate elasticity values for all sectors. As for sectors 1, 3, 4, 13 to 18 no elasticity values are available, the corresponding sectoral values from the standard specification (Case 0) are used.

Table 10-8: Sectoral values of upper-level Armington elasticities in row's import demand.

	Case 0: Standard version of GEM-E3	Case 1: Halved values	Case 2: Doubled values	Case 3: U.S. 'best guess' estimates *	Case 4: U.S. econometric estimates **
Agriculture	1.40	0.70	2.80	1.40	1.40
Coal	0.60	0.30	1.20	2.36	7.12
Crude oil and oil products	0.60	0.30	1.20	2.36	-0.34
Natural gas	0.60	0.30	1.20	0.60	0.60
Electricity	0.60	0.30	1.20	0.60	0.60
Ferrous, non-ferrous ore and metals	2.20	1.10	4.40	1.44	2.44
Chemical products	2.20	1.10	4.40	2.61	9.40
Other energy intensive industries	2.20	1.10	4.40	2.91	1.78
Electrical goods	2.20	1.10	4.40	2.11	7.46
Transport equipment	2.20	1.10	4.40	3.59	2.01
Other equipment goods industries	2.20	1.10	4.40	1.07	3.20
Consumer goods industries	2.50	1.25	5.00	2.07	2.65
Building and construction	1.40	0.70	2.80	1.40	1.40
Telecommunication services	1.40	0.70	2.80	1.40	1.40
Transports	2.20	1.10	4.40	2.20	2.20
Credit and insurance	1.40	0.70	2.80	1.40	1.40
Other market services	1.40	0.70	2.80	1.40	1.40
Non-market services	0.60	0.30	1.20	0.60	0.60

* Based upon 'best guess' U.S. estimates constructed by Shiells et al. (1986), weighted by 1988 import shares of RoW.

** Based upon U.S. econometric estimates of sector-specific substitution elasticities provided by Shiells et al. (1986), weighted by 1988 import shares of RoW.

10.2.5.1.2   **Simulation Results**

In Table 10-9, simulation results of the double dividend scenario in terms of macroeconomic aggregates are listed for the various cases defined in Table 10-8.

The variations in elasticity values have some impact on results, but, all in all, the percentage change of quantities, related to the standard case, lies within a range of ±0.5 percentage points. The sensitivity of economic welfare to alternative parameter values is not very high as well. Obviously, the EU gains the more from the double dividend policy in terms of economic welfare the less the Armington elasticity values in foreign import demand are, i.e. the less the foreign sector reacts to increasing production prices in the EU economy.

In Case 1, where values of the sectoral upper-level Armington elasticities are halved, RoW shows less strong reactions to an increase in EU export prices. While in the standard case exports fall down by -1.02%, they are reduced only by -0.92% in the case of halved elasticity values. This reflects the lower degree of substitutability between domestic and foreign production in RoW's import demand. However, at a sectoral level exports develop in different ways. Whereas for some sectors exports are less reduced compared to the standard case, for some other branches (e.g. electrical goods, equipment and consumer goods industries, transports and both service sectors) they show a higher reduction rate, or less growth rates respectively (fossil fuel sectors). The increase in world demand for EU exports in Case 1 compared to Case 0 is the main reason for a comparatively lower drop in GDP deflator. As exports, investment and consumption settle down (all at a higher level in Case 1 than in Case 0), production and gross domestic product are higher as well (-0.56% reduction of production in Case 1 instead of -0.57% in Case 0, -0.03% reduction of gross domestic product instead of -0.04%). Thus, energy consumption decreases also by a lower rate (-6.18% compared with -6.21%) and employment rises by additional 25.000 persons.

The mechanisms change direction if doubled upper-level elasticity values are introduced (Case 2). The RoW's reactions to an increase in relative prices are now stronger than in the standard case. Consequently, EU exports go down more heavily (by -1.08%). Due to diminished foreign demand, EU prices, expressed by the GDP deflator, go down to a greater extent and imports are reduced more heavily. Overall, gross domestic product and production decrease more.

While, the results in Case 3 lie like Case 0 somewhere between the two extremes, the halved and the doubled-value case, Case 4 causes stronger impacts on exports and imports. The latter is characterised by very high elasticity values for some sectors, e.g. the coal sector, chemical and electrical goods industry. All in all, aggregated exports of the EU drops by the highest percentage rate in this case, compared to all other cases. The relatively high reduction of aggregate demand expressed by a smaller increase in consumption and a greater decrease in investment and export results in a higher percentage reduction of the GDP deflator. This, in turn, leads to a higher drop in imports.

Table 10-9: Co-ordinated double dividend policyvariation of upper-level Armington elasticities in RoW's import demand.

Macroeconomic aggregates for EU-14					
	Case 0: Standard version of GEM-E3	Case 1: Halved values	Case 2: Doubled values	Case 3: U.S. 'best guess' estimates	Case 4: U.S. econometric estimates
Gross domestic product	-0.04%	-0.03%	-0.05%	-0.04%	-0.04%
Employment*	780	805	757	769	793
Production	-0.57%	-0.56%	-0.59%	-0.57%	-0.58%
Private investment	-0.18%	-0.15%	-0.20%	-0.18%	-0.19%
Private consumption	0.21%	0.29%	0.15%	0.22%	0.16%
Domestic demand	-0.56%	-0.52%	-0.58%	-0.55%	-0.58%
Exports in volume	-1.02%	-0.92%	-1.08%	-0.98%	-1.12%
Imports in volume	-1.46%	-0.99%	-1.73%	-1.33%	-1.80%
Intra trade in the EU	-1.20%	-1.26%	-1.17%	-1.20%	-1.21%
Energy consumption in volume	-6.21%	-6.18%	-6.23%	-6.21%	-6.26%
Consumers' price index	1.19%	1.65%	0.93%	1.27%	0.95%
GDP deflator in factor prices	-0.74%	-0.27%	-0.98%	-0.63%	-1.01%
Current account as % of GDP***	0.08	0.09	0.08	0.09	0.06
Equivalent variation of total welfare					
Economic welfare**	0.23%	0.27%	0.20%	0.24%	0.19%
* in thousand employed persons					
** as percent of GDP					
*** absolute difference from baseline					

(Numbers indicate percent changes from baseline except if defined otherwise)

10.2.5.2 Variation of Armington Elasticities: EU Countries

10.2.5.2.1   Specifications of Armington Elasticities

No econometric estimates of sector- and country-specific substitution elasticities for EU countries are available in the literature. Thus, in this section the required set of Armington elasticities for the 14 EU countries is generated following a procedure proposed by Harrison et al. (1991, p. 100). The procedure takes place in three steps.

1. Starting point are sector-specific 'best guess' upper-level Armington elasticities for the U.S. presented in Shiells et al. (1986). Using country-specific import weights (drawn from 1993 data[50]) for each country an *average Armington elasticity of substitution* σ_x^{av} is calculated.

---

[50] 1993 International Trade Statistics Yearbook, Vol. 1, Trade by Country, United Nations, New York.

2. The country-specific elasticities σ_x^{av} are then compared with country-specific Armington elasticities (σ_x^{inf}) that are inferred from country-specific import price elasticities (ε) and from import shares. Whereas the national import price elasticities are taken from the empirical trade literature (Stern et al. 1976), the import shares are calculated from the equilibrium benchmark data set.

3. Finally, we re-scale for each country the sector-specific elasticities so that the aggregated, import-weighted elasticity σ_x^{av} is equal to the country-specific elasticity σ_x^{inf}, which is derived from the national import price elasticity.

While step 1 and step 3 are more or less self-evident, some comments should be made on the derivation of the national Armington elasticities from literature-based import price elasticities (step 2).

Table 10-10: Country-specific price and substitution ·elasticities of import demand for different shares of non competitive imports.

| | e * | s_x^{av} ** | s_x^{inf} *** | | |
			(IMNC/IM=0)	(IMNC/IM=0.5)	(IMNC/IM=0.8)
Austria	-1.32	2.13	1.88	4.57	10.48
Belgium	-0.83	2.13	1.67	5.03	6.53
Germany	-0.88	2.12	1.09	2.90	6.09
Denmark	-1.05	1.99	1.53	2.61	-10.31
Finland	-0.5	2.37	0.62	2.97	3.46
France	-1.08	1.63	1.31	2.36	7.31
Greece	-1.03	2.15	1.04	2.10	5.24
Ireland	-1.37	1.95	1.62	2.65	8.94
Italy	-1.03	2.01	1.77	6.39	8.57
Netherlands	-0.68	2.03	1.20	3.32	5.63
Portugal	-1.03	1.92	1.33	3.05	7.52
Spain	-1.03	2.03	1.21	2.63	6.68
Sweden	-0.79	2.06	0.80	1.38	1.99
United Kingdom	-0.65	1.93	0.66	1.17	1.83

Obviously, the procedure proposed is faced with some problems, which arise from the existence of non tradable sectors and non competitive imports. Both import demands of non traded and non competitive commodities are excluded from the Armington assumption. It is assumed that they are determined not by price relations but by the domestic production level and institutional settings, such as supply contracts. As national import price elasticities, taken from the literature, normally refer to the national aggregate of import demand (aggregating all sectors), they may provide a distorted picture of Armington elasticities. However, this problem is less important here. Fortunately, in the GEM-E3 model the national shares of imports of non tradable goods in total imports are low and by a majority below 5%. Thus, the literature-based import price elasticity values are reasonable approximates for the price elasticity of import demand of tradable goods in the GEM-E3 model.

More importance should be attached to the problem arising from non competitive imports. Given the same import price elasticity value, the share of non competitive imports assumed influences the inferred Armington elasticity values σ_x^{inf} decisively.

A variation of the share of non competitive imports in total imports of tradable goods leads to different values of Armington elasticities. In summary, one can say that the higher the share of non competitive imports, the higher the Armington elasticity which corresponds to a given import price elasticity. In the GEM-E3 model the shares of non competitive imports are set equal to 0.5 for all countries and all sectors.

Table below reports the values of the upper-level Armington elasticity for which sensitivity analysis is performed. As in the previous section, the case of doubled and halved elasticity values are tested.

Table 10-11: Sectoral values of upper-level Armington elasticities in RoW's import demand.

	Case 0: Standard version of GEM-E3	Case 1: Halved values	Case 2: Doubled values	Case 3: U.S. 'best guess' estimates
Agriculture	1.2	0.60	2.40	
Crude oil and oil products	0.6	0.30	1.20	Country- and sector-
Ferrous, non-ferrous ore and metals	1.5	0.75	3.00	specific values
Chemical products	1.5	0.75	3.00	
Other energy intensive industries	1.5	0.75	3.00	(for sectors 3, 7- 10:
Electrical goods	1.5	0.75	3.00	values as calculated from
Transport equipment	1.5	0.75	3.00	'best guess' estimates
Other equipment goods industries	1.5	0.75	3.00	presented in Shiells et al. (1986);
Consumer goods industries	1.7	0.85	3.40	for sectors 1, 6, 11-17:
Telecommunication services	0.6	0.30	1.20	values as in standard version)
Transports	1.2	0.60	2.40	
Credit and insurance	0.6	0.30	1.20	
Other market services	0.6	0.30	1.20	

10.2.5.2.2 Simulation Results

The four cases of parameter choice differ only slightly with respect to macroeconomic impacts. Differences arise mainly in trade flows and price indices. All other macroeconomic variables show only marginal changes. As consumption and employment, or leisure respectively, remains nearly constant, economic welfare also varies scarcely.

The interpretation of results starts with examining the pure effects of a variation of Armington elasticities.

In Case 1, Armington elasticity values in the aggregate import demand of all EU countries are halved, i.e. substitution possibilities between domestic production and imports are more restricted for all EU countries. For instance, in Case 1 a policy-induced price increase in European domestic supply will induce a lower substitution effect than in the standard version, i.e. import demand for tradable goods will be expanded to a lower extent than in the standard case. This, however, implies that in Case 1 (relatively expensive) domestic production has a relatively bigger share in overall EU domestic supply. The pure substitution effect leads, in Case 1, to relatively higher prices. The latter is expressed by a decrease of the GDP deflator by -0.72% (compared to -0.74% in Case 0).

In Case 2, we have to argue the other way round. Doubling the Armington elasticities increases the substitution effect, i.e. a shift in the price relation of domestic supply and imports results in a comparatively higher demand for imports. Now, domestic production constitutes a lower part in domestic supply compared with both Case 0 and Case 1. This, in turn, results in a decrease in domestic prices. To conclude, the price level is at its highest in Case 1 and at its lowest in Case 2. Case 0 lies in between.

In the standard model, the double dividend policy results in a decrease in exports and imports and in a drop in the GDP deflator (resulting from a decrease in aggregate demand).

As just mentioned in the case of halved Armington elasticity values (Case 1) the drop in the GDP deflator is less significant, i.e. prices are higher. This exactly reflects the cost effects of a lower degree of substitution of European producers and consumers. Due to comparably higher prices, EU exports are reduced to a greater extent. Whereas exports fall by -1.02% in the standard version, they fall by -1.05% in Case 1. The greater percentage reduction of imports can be explained by reduced substitution possibilities in Case 1, i.e. import demand increases less in response to an increase in domestic production prices.

In the case of doubled Armington elasticities (Case 2) domestic prices are lower. Thus, exports are reduced to a lower extent (by -0.97% compared to -1.02% in Case 0). Imports decrease to a lower extent as well (by -1.42% compared to -1.46% in Case 0) due to the higher substitution possibilities given by doubled elasticity values.

Table 10-12: Co-ordinated double dividend policy variation of upper-level Armington elasticities in import demand of EU countries.

	Macroeconomic Aggregates for EU-14			
	Case 0: Standard version of GEM-E3	Case 1: Halved values	Case 2: Doubled values	Case 3: U.S. 'best guess' estimates
Gross Domestic Product	-0.04%	-0.04%	-0.04%	-0.04%
Employment*	780	784	774	775
Production	-0.57%	-0.57%	-0.58%	-0.58%
Private Investment	-0.18%	-0.17%	-0.18%	-0.18%
Private Consumption	0.21%	0.21%	0.20%	0.20%
Domestic Demand	-0.56%	-0.55%	-0.56%	-0.56%
Exports in volume	-1.02%	-1.05%	-0.97%	-1.02%
Imports in volume	-1.46%	-1.48%	-1.42%	-1.45%
Intra trade in the EU	-1.20%	-1.25%	-1.11%	-1.14%
Energy consumption in volume	-6.21%	-6.21%	-6.22%	-6.25%
Consumers' price index	1.19%	1.21%	1.16%	1.18%
GDP deflator in factor prices	-0.74%	-0.72%	-0.77%	-0.73%
Current account as % of GDP***	0.08	0.08	0.08	0.08
Equivalent Variation of total welfare				
Economic welfare**	0.23%	0.23%	0.22%	0.23%
* in thousand employed persons ** as percent of GDP *** absolute difference from baseline				

(Numbers indicate percent changes from baseline except if defined otherwise)

10.2.6   Conclusion

The specification of the world closure, i.e. the way of closing the domestic economy model by incorporating the external sector, is a crucial component for those models, in which production and consumption is not specified endogenously for all countries. Here, reasonable assumptions concerning the behaviour of the RoW have to be made, often in combination with a balance-of-payments constraint.

The closure rule incorporated in the GEM-E3 model is advantageous in empirical application as it, for instance, avoids complete specialisation in production, allows for modelling of intra-industrial trade and includes non traded and traded goods. In particular, intra-EU trade activities that account for around 60% of the whole EU trade are modelled realistically as they depend on an endogenous international price system. But even if the GEM-E3 model takes a mainly European perspective, the specification of the foreign sector has a great deal of influence. Relaxing the assumption of fixed prices for exports of the RoW facing the EU as a whole seems to be important. Furthermore, a better link between both economies, the EU economy and the economy of the RoW, should be considered.

In this chapter, two main changes in the foreign trade specification have been proposed and tested. The basis is a simulation of a co-ordinated double dividend scenario. The first change refers to the RoW's export supply function in which a constant finite price elasticity has been introduced. The second change concerns the RoW's import demand function in which a RoW's activity variable was incorporated. In summary, the impact in terms of economic welfare and changes in macroeconomic variables is noteworthy for the former case while no substantial changes could be observed for the latter case.

Additionally, the sensitivity of the GEM-E3 model to variations in key parameter values such as the upper-level Armington elasticity has been analysed. Results indicate that the model can be interpreted as quite robust to parameter changes. Thus, exact econometric estimations of upper-level Armington elasticities are undoubtedly an important issue, but should not be a priority for future research.

To conclude, a comprehensive solution to the problems outlined above will be best tackled by extending the regional scope of the GEM-E3 model towards a global model with an endogenous representation of the behaviour of agents of the RoW. Thus, future research on the GEM-E3 model will concentrate on a better understanding of production and consumption activities in the RoW as a whole and on a further disaggregation in several major trading blocks.

11 Conclusions

11.1 Emission Abatement and Sustainable Development

As shown by the analysis in the previous chapters, reducing CO_2 emissions can invoke serious global changes in the economic environment of the European Union. Because of its all-pervasive character CO_2 policy affects most economic sectors and agents. Assuming, as it is standard in the CGE literature, that the initial situation is one that is Pareto optimal, then the policy can only result in welfare reductions (at least in economic terms, i.e. excluding environmental benefits). If on the other hand unexploited resources are assumed to exist in the baseline scenario, such as for example with respect to the rest-of-the-world, or in the labour market or unexploited technological options, then the environmental policy might with suitable design use this potential to achieve low, zero or even negative economic policies.

The goal of the analysis in the previous chapters was to identify such potential deviations from Pareto optimality and use them to reduce the cost of reaching alternative emission abatement targets. Three main potential deviations where analysed in detail.

- *Distortionary taxation in the labour market and in investments:* If the current system of labour and investment taxation are inefficient, then by using the revenues of a carbon tax, one could reduce the inefficiencies, releasing economic growth potential. GEM-E3 results have shown that it is indeed possible to get employment gains or investment increases in this way. However the impact on GDP is always negative and the impact on consumer's welfare crucially depending on the labour market assumptions, is usually also negative.

- *Terms-of-trade potential:* Alternative assumptions on the degree of exposure to foreign competition may imply that there may exist the potential to gain from terms-of-trade changes with respect to the rest of the world, or alternatively that any deterioration in competitiveness may lead to very negative results for some sectors. The mainstream results of GEM-E3 allow for limited such gains, showing that there may be a limited possibility for gains vis-à-vis the rest of the world, while sensitivity tests evaluate the stability of this assumption.

- *Technological possibilities:* One might assume that there exists in the baseline energy-saving technology options that although not significantly more expensive than existing ones, they are not adopted by firms and households. Such an assumption requires modifying the model to explicitly incorporate energy-saving investments. The exact specification then of their costs, defines the shape of the marginal cost curve and the low-cost abatement possibilities. A variant of the GEM-E3 model was developed to examine the impact of such assumptions. There was assumed that there is only limited potential for low-cost energy-saving investments. The results of this model verify the general conclusions drawn from the main model version but also illustrate the assumptions with which the abatement cost could be much lowered and technology-driven sustainable development can be accomplished.

11.2 Policy Instruments

GEM-E3 can model several aspects of environmental policy. It incorporates taxes and pollution permits that can be either unilateral or Europe-wide. It can set an emission reduction target and compute the corresponding tax or pollution permit price consistent with the goal.

Concerning accompanying policies a number of possibilities exist: in the case where a tax is imposed, revenues are initially given to the government.

Of course mixed policies, such as for example a tax in some sectors or in the households and pollution permits in others such as for example in energy intensive industries is also possible.

11.3 Policy Evaluation Criteria and Main Results

The marginal abatement cost of emission reduction is generally significant. Due to the nature of the GEM-E3 model, where energy efficiency gains are modelled through neo-classical production functions there is no inherent assumption for free or very cost-effective investment in energy saving. In the model variant with explicit energy-saving investments such an assumption could have been made but was not adopted for comparability of results.

The overall marginal abatement cost curves follows then a similar slope in all simulations, although the level varies depending on the type of accompanying policies adopted. For example it is highest in the case of pollution permits and lowest is the case of subsidisation of investment costs. But this does not in any way signify that the former policy is more effective than the latter. As GEM-E3 is a general equilibrium model representing the whole economic system, the marginal abatement cost per se, is not a very useful concept. For instance, one reason why the cost is highest in the case of pollution permits is that because of its effectiveness as a policy instrument, pollution permits lead to a preservation of activity and so naturally to higher abatement costs.

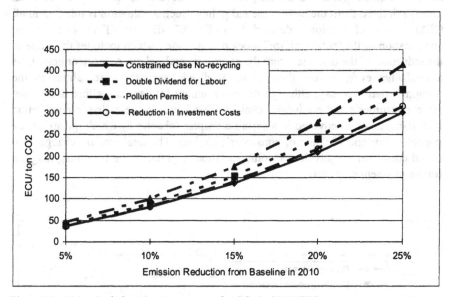

Figure 11-1: Marginal abatement cost curve for CO_2 in 2010 (EU).

Much more important insights can be gained by looking at the macroeconomic indicators, such as GDP, employment, and competitiveness. All these are provided in the relevant chapters of this book and vary from scenario to scenario depending on the impact of the accompanying policies.

As standard in a CGE model, GEM-E3 by directly reflecting economic theory provides a unique criterion for evaluating a policy: consumer's welfare. By mapping economic theory, GEM-E3 constructs from observable quantities (consumption and employment) an unobservable construct, i.e. the consumer's welfare. It is exactly the changes in this welfare that are the absolute criterion for evaluating a policy.

Two problems exist in this respect: although the welfare function in GEM-E3 is derived in a standard way found in economic literature, there remains the problem of valuation of unemployment or leisure. On one hand, for a person that is currently employed, working more can be seen as, ceteris paribus, a reduction in welfare while for an unemployed person finding a job is certainly welfare-increasing. This problem is still unanswered in the theory of welfare economics and no single satisfactory answer exists. In the presentation of the results we have chosen to show both welfare indicators (i.e. welfare-increasing and welfare-decreasing unemployment).

A second problem comes for the fact that changes in environmental quality should result in welfare gains. To do that, one should be able to quantify the external benefits derived from the environmental policy. Such a valuation is inherent in all GEM-E3 model versions, derived from EXTERNE data. This leads to the computation of the "total welfare" index of the model, which includes the value of the reduction of the damages from the environmental policy. A shortcoming of the analysis however, is that there is no feedback from the externalities to the economy, that is to say, although the environment influences welfare it does not affect consumer's and producer's choices. This issue, the so-called internalisation of external costs, is covered in a separate chapter (chapter 8) which quantifies the impact of this specification of consumer's choices. The total welfare change, also called consumer's equivalent income variation is given in the following diagram for the different scenarios.

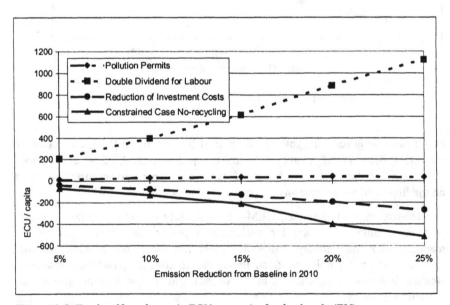

Figure 11-2: Total welfare change in ECU per capita for the decade (EU).

Concerning the double dividend issue, GEM-E3 results consistently draw results that are in favour of the "weak" double dividend hypothesis and against its "strong" variants. The environmental policy is likely to negatively affect consumer's welfare, but can positively affect specific targets: for example employment gains can be effected (by using the revenues from the carbon tax to subsidise labour costs) or investments can be augmented (when tax revenues are used to subsidise investment costs). Also in most cases total equivalent variation can be positive, i.e. environmental benefits can outweigh the costs. Total activity however, is generally found reduced, while competitiveness also usually deteriorates, raising question about the sustainability of the gains.

Of importance are also the distributional effects of the policy concerning the different sectors, but again the exact effects depend on the accompanying policies. In "labour recycling" cases, demand is re-oriented towards the consumer goods industries and away from energy intensive and equipment. Such a structural shift has negative implications on investments and probably important long-term implications not captured by the model.

The quantitative nature of the results was usually all verified in the sensitivity tests performed, but the exact figures vary significantly depending on the accompanying policy and the assumption about the possibility of adjustment of the economic agents. If for example we had adopted the postulate that energy intensive industries are already on the limit of their possibilities for energy conservation or that they operate in an international environment which is very competitive, then the results would be more negative. Under extreme assumptions, even the employment dividend disappears.

If, on the other hand, there exists in the economy the potential to achieve significant efficiency gains in energy use at a moderate cost, then the negative impact of environmental constraints on the economy may be lower. In this case additional benefits from terms-of-trade effects may further reduce the cost of the policy. In this case, it is possible that there is no net economic cost for the policy, supporting the "strong" double dividend hypothesis. But this requires rather extreme assumptions (very low exposure to foreign competition, high potential for cheap energy-saving investments, high flexibility of the labour market).

The distinction between pollution permits and taxes is irrelevant in GEM-E3. Since the economic agents operate under perfect information and total resources are allocated so as to reduce total costs, there is no theoretical distinction between taxes and permits. In pervious chapters this equivalence has been demonstrated theoretically, in the sense that taxes can be thought of as permits with a specific allocation of resources and vice versa.

So taxes vs. permits is a question that cannot be answered with GEM-E3, but different burden sharing schemes can be evaluated. The environmental target was always set at the EU-14 level, and the allocation of effort was distributed across the countries so as to minimise total costs. The tax burden (or equivalent permit allocation) by economic agent and country is however arbitrary. In the main model applications simplifying assumptions where made on both issues: on the permits simulations the initial endowment follows the "grandfathering" principle, i.e. permits proportional to the base year emission by agent; on the tax case the revenues where recycled within the country given to labour or investments. The chapter on burden sharing modifies this assumption and shows its importance. Under different permit allocation schemes, the results vary significantly with some countries gaining and some countries losing. There of course no single criterion exists and the question of equity is raised. To what extent some countries may take more burden than others? Cohesion targets may imply for example that the burden per capita or per unit of GDP might be equal.

Sensitivity tests were also done on a similar issue, namely whether it is worth to exempt some sectors or derogate action in some countries for some years. Exemptions where studied for the energy intensive industries and it was found that indeed these sectors suffer much less. As a matter of fact, it may even be possible that, in a first approximation, these sectors might also benefit (e.g. from lower labour costs, from subsidised investments or from selling permits). However the additional burden facing the other sectors (because the CO_2 target is given) has negative implications for the economy as a whole, so that indirectly energy intensive industries suffer as well. The final outcome of the simulations is that the overall effectiveness of the policy when energy intensive industries are exempted gets lower.

Derogations are again an issue that can only partly be covered by GEM-E3. Due to the fact that expectations in GEM-E3 are backward looking the fact that some countries face the tax, will not alter the policy of companies operating in countries that do not have the cost. The only effect that remains is then through trade, where lower prices in countries with no environmental constraint will augment their market, but only for the period of the derogation. When these countries finally start imposing a constraint on emissions then they have to undergo a transitory period, which the other countries have now overcome. So, this policy is likely to lead to some short term gains followed by longer term losses. Again however the overall impact is rather limited. Europe-wide effectiveness is again lower than in the main scenario. The message from the model, is that derogations are not effective from an economic point of view and should only be selected on the base of other considerations (cohesion targets) and then only if accompanied by some specific policy to make the period of exemption more fruitful.

12 APPENDIX: Brief Description of Models Used in the Study[51]

The GEM-E3 model has been developed within a series of collaborative research projects partially funded by the European Commission, Joule Programme of DG XII. Pierre Valette and Huguette Laval of Unit F1 of DG XII have led scientific supervision from the European Commission.

12.1 GEM-E3

12.1.1 Introduction

GEM-E3 is a computable general equilibrium model, for the European Union member states, that provides details on the macro-economy and its interaction with the environment and the energy system. It is an empirical, large-scale model, written entirely in structural form. The model computes the equilibrium prices of goods, services, labour and capital that simultaneously clear all markets under the Walras law. In brief, the model can be characterised as follows:

- it is a multi-country model, treating separately each EU-15 member-state and linking them through endogenous trade of goods and services;

- it includes multiple industrial sectors and economic agents, allowing the consistent evaluation of distributional effects of policies;

- it is a multi-period model, involving dynamics of capital accumulation and technology progress, stock and flow relationships and backward looking expectations.

---

[51] Detailed manuals of the models contained in this Appendix can be found in the full report under the project JOS3-CT95-0008

In addition, the model covers the major aspects of public finance including all substantial taxes, social policy subsidies, public expenditures and deficit financing, as well as policy instruments specific for the environment/energy system. A financial/monetary sub-model complements the real side of the model and operates as an overall closure, following the IS-LM methodology. Therefore, this macro-closure is combined with a micro-economic representation of agents' behaviour and market clearing.

The model determines the optimum balance of energy demand and supply, atmospheric emissions and pollution abatement (for acid rain emissions), simultaneously with the optimising behaviour of agents and the fulfilment of the overall equilibrium conditions. In this sense, the model analyses the interactions between the economy, the energy and the environment systems.

The results of the model include projections of full input-output tables by country, national accounts, employment, capital, monetary and financial flows, balance of payments, public finance and revenues, household consumption, energy use and supply, and atmospheric emissions. The computation of equilibrium is simultaneous for all domestic markets of all EU-15 countries and foreign trade links.

A major aim of the model in supporting policy analysis, is the consistent evaluation of distributional effects, across countries, economic sectors and agents. The burden sharing aspects of energy supply and environmental protection are fully analysed, while ensuring that the European economy remains at a general equilibrium condition.

12.1.2 Model Design Principles

The *GEM-E3* model includes a detailed representation of production structures and consumption patterns, with fully flexible coefficients, as in the D. W. Jorgenson's tradition. It also integrates energy and environment within production costing and the Input-Output structure.

The model is largely inspired by the computable general equilibrium models that are extended with an IS-LM closure. The model implements the so-called macro-micro approach that combines a micro-economic representation of demand and supply behaviour with a macro-economic consistent framework.

In addition, it formulates endogenous international trade and detailed treatment of taxation, as in other trade-oriented general equilibrium models.

The model is based on a Social Accounting Matrix framework and is calibrated for a base year, as in World Bank models. The model runs dynamically by solving a system of simultaneous non linear equations in each period.

12.1.3 Model Components

Domestic production is defined by sector following the sectoral decomposition of the Input-Output table. It is assumed that each sector produces a single good, following a constant return of scale technology. It is also assumed that perfect competition conditions prevail in all markets for goods. The sectoral firm decides, under profit maximisation, its supply of good or service given its selling price and the prices of production inputs. The firm can change its stock of capital only in the next year, by investing in the current year. Since the stock of capital is fixed within the current year, the supply curve of domestic goods is upwards sloping and exhibits decreasing return of scale. All production inputs are considered as production factors. The firm adjusts flexibly the entire mix of production inputs.

The desired stock of capital for the next year is derived from profit maximisation seeking to achieve an optimal level of long run rate of return of capital, given expectations about future user's cost of capital. The optimal long-run rate of return of capital is derived according to Ando-Modigliani formula involving the real interest rate augmented by the depreciation rate. Sectoral investment is therefore obtained through partial adjustment of current to desired stock of capital.

The behaviour of the representative household is derived on an inter-temporal model. The model has to determine household's decision regarding his labour supply and the allocation of revenues into consumption, investment and savings. This decision is conceived for a given wage rate (derived from the labour market), interest and discount rates. Labour supply depends on the allocation of fixed time resources of households into leisure and work. The computed demand for leisure indirectly determines total expenditure and savings of households. Given households' total spending in investment (dwellings) and consumption of goods, the model determines the derived demand for goods and services. The allocation mechanism is flexible, price dependent and considers durable and non durable goods. Non durable goods are associated to consumption purposes (food, culture etc.). Durable goods include cars, heating systems and electric appliances, and their use involves demand for non durable goods, for example energy.

Government behaviour is largely exogenous. The government demands for consumption goods and services and for capital goods to form public investment. Following the SAM definitions, the model distinguishes between several categories of public revenues, depending on a variety of taxation and other policy instruments. Public transfers to economic agents are also represented according to the SAM definitions. For example, government allocates transfers for social policy and receives social security contributions.

The demand of products by the consumers, the producers and the public sector (for consumption and investment) constitutes total domestic demand.

This domestic demand is allocated between domestic products and imported products, following the Armington specification. In this specification, sectors and consumers use, under cost minimisation, a composite commodity that combines domestically produced and imported goods, which are considered as imperfect substitutes. The minimum unit cost of the composite good determines its selling price. This is formulated through a CES unit cost function, involving the selling price of the domestic good, which is determined by goods market equilibrium, and the price of imported goods, which is taken as an average over countries of origin. By applying Shephard's lemma, we derive total demand for domestically produced goods and for imported goods. In *GEM-E3* imports are further allocated by country of origin, depending on their relative export prices. Dual unit-cost formulation is used throughout.

The supplier of domestically produced goods faces two markets: the domestic and the foreign ones. Therefore, in order to maximise his profits, we assume that he can apply different tariffs according to the nature of these two markets. We introduce an export supply function to reflect the producer's decision on the optimal mix of goods offered to the domestic market and goods offered to the world market, following a Constant Elasticity of Transformation (CET) function operating under profit maximisation.

The model is not covering the whole planet and thus the behaviour of the rest of the world (ROW) is left exogenous: imports demanded by the rest of the world depend on export prices set up by the European Union countries, while exports from the rest of the world to the EU are sold at an exogenous price. The imports demanded by the ROW are flexibly satisfied by exports originating from the European Union (EU) countries. The latter consider the profitability of exporting to the ROW, exporting to EU or addressing the goods to their domestic markets. Via these profitability considerations, the EU countries set their export prices, as mentioned above. Within the European Union, exports are considered homogeneous. This means that the producer sets a single export price for European Union countries and another export price for the rest of the world.

Imports demanded by the EU countries from the ROW are supplied by the latter flexibly. However, the EU countries consider the optimal allocation of their total imports over the countries of origin, according to the relative import prices. Each country buys imports at the prices set by the supplying countries following their export supply behaviour. Of course, the supplying countries may gain or loose market shares according to their price setting. When importing, the EU countries compute an index of mean import price according to their optimal allocation by country of origin. This mean import price is then compared to the domestic prices in order to allocate demand between imports and domestic production (this corresponds to an Armington assumption).

The model verifies analytically that the balance of the trade matrix in value and the global Walras law are verified in all cases. A trade flow from one country to another matches, by construction, the inverse flow. The model ensures this symmetry in volume, value, and deflator. Income flows between agents, following the SAM definitions as mentioned before, and the market equilibrium conditions complete the model. The equilibrium conditions in the markets of goods serves to determine domestic production prices.

For the labour market it is postulated that wage flexibility ensures full employment. On the demand side we have the labour demanded by firms (as derived from their production behaviour), while on the supply side we have the total available time resources of the households minus the households' desire for leisure (which is derived from the maximisation of their utility function). The equilibrium condition serves to compute the wage rate.

At the equilibrium point, the economic agents either achieve maximum profit (equal to zero) or completely use their budget constraint. The model then verifies the Walras law at the global level.

The model evaluates the energy-related emissions of CO_2, NO_x and SO_2, VOC and PM as a function of the energy consumption and the abatement level per sector. For SO_2 and NO_x we specify abatement costs which will increase the cost of using pollution intensive inputs. The cost of energy, as considered in optimising behaviour of producers and consumers, consists of the cost of acquiring energy inputs, the costs of abatement equipment and the costs (or revenues) from transacting CO_2 emission rights. Therefore, environmental decisions of economic agents (regarding energy, abatement and pollution permits) are taken simultaneously with non environmental (that is economic) decisions.

12.1.4 Nomenclature

The latest available version of the GEM-E3 model (version 2.0) represents:

- all EU-15 countries, separately and linked
- 18 products and sectors:
    1. agriculture
    2. solid fuels
    3. liquid fuels
    4. natural gas
    5. electricity
    6. ferrous and non ferrous metals
    7. chemical industry
    8. other energy intensive industries
    9. electrical goods

10. transport equipment

11. other equipment goods

12. consumer goods

13. building and construction

14. telecommunication services

15. transports

16. service of credit and insurance institutions

17. market services

18. non market services

- 4 economic agents: households; firms; government; foreign sector
- several government revenue and income flow categories:

    direct taxation, indirect and VAT taxation

    energy and environmental taxation

    property taxes, capital taxes

    social security, social benefits

    subsidies (production and exports)

    import duties and foreign sector transfers

    revenues from government enterprises

- 13 household expenditure categories:

    9 non durable consumption categories (food, culture, health, electricity, gas, motor fuels, other fuels, transport, house)

    3 durable consumption categories (cars, heating systems, electrical appliances)

    investment in dwellings

- 2 primary production factors: labour; capital
- 6 pollutants: CO_2, SO_2, NO_x, VOC, and PM

12.1.5   Results from GEM-E3

- Dynamic annual projections in volume, value and deflator of national accounts by country.

- Full Input-Output tables by country and for EU-15 as a whole, for the 18 sectors.

- Distribution of income and transfers in the form of a social accounting matrix by country.

- Employment, capital, investment by country and sector.

- Monetary/financial flows and current account by country.

- Atmospheric emissions, pollution abatement capital, purchase of pollution permits and damages.

- Consumption matrix by product and investment matrix by ownership branch.

- Public finance, tax incidence and revenues by country.

- Full trade matrix for EU-15 and the rest of the World.

- The current version of GEM-E3 evaluates about 60,000 simultaneous equations per year, and follows a time-forward path.

- The solution algorithm uses a combination of Gauss-Seidel and Newton Successive Over-Relaxation methods.

- The model uses the Solver/NTUA modelling software (version 1.9) and operates in MS-Windows.

List of Figures

200

List of Tables

202

References

Adams, R. et al. (1993), "Sequestering Carbon on Agricultural Land: Social Cost and Impacts on Timber Markets," Contemporary Policy Issues, XI (1), p. 76-87.

Asian Development Bank (1993), "National Response Strategy for Global Climate Change", People's Republic of China.

Atkinson, S. E., Tietenberg, T. H. (1982), "The Empirical Properties of Two Classes of Designs for Transferable Discharge Permit Markets", in Journal of Environmental Economics and Management, Vol. 9, p. 101-121.

Auerbach, A.J. (1985), "The Theory of Excess Burden and Optimal Taxation", in Auerbach A.J., Feldstein M., (Eds.) Handbook of Public Economics, Vol. I, p. 61-127.

Bach, S., Kohlhaas, M., Praetorius, B. (1996), "Economic Effects of an Ecological Tax Reform", in Hohmeyer, O., Ottinger, R.L., Rennings, K. (Eds.) *Social Costs and Sustainability*, Berlin Heidelberg, p. 451-464.

Baldwin, R. (1995), "Does Sustainability Require Growth", in Goldin, I. and Winters, A. (Eds.) *The Economics of Sustainable Development*, Cambridge, Cambridge University Press.

Ballard, C.L. and Medema, S.G. (1993), "The Marginal Efficiency Effects of Taxes and Subsidies in *The Presence of Externalities*, Journal of Public Economics, Vol. 52, p. 199-216.

Barker, T. (1994), "Taxing Pollution instead of Jobs: Towards more Employment without more Inflation through Fiscal Reform in the UK", Forthcoming in Carraro C. and Siniscalco D., (Eds.) *Environmental Fiscal Reform and Unemployment*, Dordrecht: Kluwer Academic Pub.

Barns, D.W. et al. (1992), "The Use of the Edmonds-Reilly Model to Model Energy-Related Greenhouse Gas Emissions", Organisation for Economic Co-operation and Development, (OECD), Economics Department Working Papers, n.113, Paris, France.

Barrett, S. (1994), "Climate Change Policy and International Trade" in "Climate Change: Policy Instruments and their Implications", Proceedings of the Tsukuba Workshop of IPCC WG III, Tsukuba, Japan, 17-20 January 1994.

Barrett, S. (1997), "International Environmental Agreements: a Game Theoretic Perspective", Paper presented at FEEM-EMF Stanford-IPCC Conference on International Environmental Agreements on Climate Change, Venice, May 6-7, 1997.

Baumol, W. J. and Oates, W.E. (1988), "The Theory of Environmental Policy", Second Edition, Cambridge.

Bergman, L. (1991), "General Equilibrium Effects of Environmental Policy: A CGE-Modelling Approach", Research Paper n.6415, Handelshgskolan, Stockholm, Sweden.

Berndt, E.R. and Wood, D.O. (1983), "Energy Price Changes and the Induced Revaluation of Durable Capital in U.S. Manufacturing", MIT Sloan School Working Paper No.1455-83.

Bernstein, P. M., Montgomery, W. D. and Rutherford, T. F. (1997), "World Economic Impacts of US Commitments to Medium Term Carbon Emissions Limits", Final Report to the American Petroleum Institute, Charles River Associates, Report No.873-06, January 1997.

Blanchard, O. and Fisher, S. (1989), Lectures on Macroeconomics, Cambridge, MIT Press.

Bloch, F. (1997), "Non Cooperative Models of Coalition Formation in Games with Spillovers", Forthcoming in Carraro, C. and Siniscalco, D. (Eds.) *New Directions in the Economic Theory of the Environment*, Cambridge University Press, Cambridge.

Bodlund, B. et al. (1989), "The Challenge of Choice: Technology Options for the Swedish Electricity Sector", in Johansson et al. (1989), in *Electricity: Efficient End-use and New Generations of Technologies, their Planning Implications*, Lund University press, Sweden.

Boehringer, C., Pahlke, A., and Rutherford, T.F. (1996), "The Prospects for a Double Dividend in Germany", First Interim Report, IER University of Stuttgart, March.

Boero, G. et al. (1991), "The Macroeconomic Consequences of Controlling Greenhouse Gases: A Survey", Department of the Environment, Environmental Economics Research Series, London, HMSO.

Boetti, M. and Botteon, M. (1994), "Environmental Policy, Employment and Competitiveness: The Role of Best Available Technologies", in *Le progrès technologique pour la compétitivité et l'emploi: le cas énergie-environment*, Offices des Publications des Communautés Européennes, Bruxelles.

Bohm, P. and Russell, C.S. (1985), "Comparative Analysis of Alternative Policy Instruments", in Kneese, A. and Sweeney, J. (Eds.) *Handbook of Natural Resource and Energy Economics*, Vol. I, The Netherlands, p. 395-460.

Bollen, J. and Gielen, A. M. (1997a), "Economic Analysis of Multilateral CO2 Emission Reduction Policies - Simulations with WORLDSCAN", Paper presented at FEEM-EMF Stanford-IPCC Conference on International Environmental Agreements on Climate Change, Venice, May 6-7, 1997.

Bonus, H. (1990), "Preis-und Mengenlösungen in der Umweltpolitik", in *Jahrbuch für Sozialwissenschaft*, Jg. 41, p. 343-358.

Bonus, H. (1994), Vergleich von Abgaben und Zertifikaten. Arbeitspapier Nr. 3 der Forschungsgruppe Umweltökonomie und Umweltmanagement an der Universität Münster.

Boone, L., Hall, S. and Kemball-Cook, D. (1992), "Endogenous Technical Progress in Fossil Fuel Demand: The Case of France", Centre for Economic Studies, Discussion Paper No.21-93.

Bound, G. and Johnson, G. (1992), "Changes in the Structure of Wages during the 80s: an Evaluation of Alternative Explanations", American Economic Review 82, 371-392.

Bovenberg, A.L. and F. Van Der Ploeg (1994), "Green Policies and Public Finance in a Small Open Economy", Scandinavian Journal of Economics, Vol. 96, p. 343-363.

Bovenberg, A.L. and Goulder, L. H. (1996), "Optimal Environmental Taxation in the Presence of Other Taxes: General Equilibrium Analysis", American Economic Review, Vol. 86, No.4, p. 985-1000.

Bovenberg, A.L. and R.A. de Mooij (1994a), "Environmental Levies and Distortionary Taxation", American Economic Review, Vol. 94/4, p. 1085-1089.

Bovenberg, A.L. and R.A. de Mooij (1994b), "Environmental Taxes and Labor-Market Distortion", European Journal of Political Economy, Vol. 10, p. 655-683.

Bovenberg, L. (1994), "Environmental Policy, Distortionary Labour Taxation and Employment: Pollution Taxes and the Double Dividend", Forthcoming in Carraro, C. and Siniscalco, D. New Directions in the Economic Theory of the Environment, Cambridge University Press, Forthcoming.

Bovenberg, L. and Goulder, L. (1993), "Integrating Environmental and Distortionary Taxes: General Equilibrium Analysis", Paper presented at the Conference on "Market Approaches to Environmental Protection", Stanford University, December 3-4, 1993.

Bovenberg, L. and R. de Mooij (1994), "Environmental Levies and Distortionary Taxation" American Economic Review, 4, 1085-1089.

Bovenberg, L. and R. Van der Ploeg (1992), "Environmental Policy, Public Finance and the Labour Market in a Second Best World", Centre for Economic Policy Research Discussion Paper No.745.

Bovenberg, L. and Van der Ploeg, R. (1993a), "Green Policies in a Small Open Economy", Centre for Economic Policy Research, Discussion Paper No.785.

Bovenberg, L. and R. Van der Ploeg (1993b), "Does a Tougher Environmental Policy Raise Unemployment? Optimal Taxation, Public Goods and Environmental Policy with Rationing of Labour Supply", Centre for Economic Policy Research Discussion Paper No.869.

Bovenberg, L. and R. Van der Ploeg (1994), "Optimal Taxation, Public Goods and Environmental Policy with Involuntary Unemployment", Paper presented at the NBER-University of Turin-FEEEM Conference on "Market Failures and Public Policy", Turin, 19-21 May 1994.

Brander, J. and Taylor, S. (1997), "International Trade Between Consumer and Conservationist Countries", Forthcoming, Resource and Energy Economics.

Breier, S. (1997), "Umweltschutzkooperationen zwischen Staat und Wirtschaft auf dem Prüfstand - Eine Untersuchung am Beispiel der Erklärung der deutschen Wirtschaft zur Klimavorsorge", in Zeitschrift für Umweltpolitik & Umweltrecht, No. 1, p. 131-142.

Brown, S., Feng, L., Kennedy, D. and Fisher, B. (1997), "The Economics and Reality of International Climate Change Policy Development", Paper presented to Australian Petroleum Production and Exploration Conference, Melbourne, 13-16 April 1997.

Brunello, G. (1996), "Labour Market Institutions and the Double Dividend Hypothesis", in Carraro, C. and Siniscalco, D. (Eds.), *Environmental Fiscal Reform and Unemployment*, Dordrecht, Kluwer Academic Pub.

Bundesverband der Deutschen Industrie, Aktualisierte Erklärung der deutschen Wirtschaft zur Klimavorsorge vom 27.03.1996.

Burniaux, J.M., Martin, J.P, Nicoletti, G. and Oliveira, J. (1991), "GREEN A Multi-Region Dynamic General Equilibrium Model for Quantifying the Costs of Curbing CO_2 Emissions: a Technical Manual", Organisation for Economic Co-operation and Development, (OECD), Department of Economics and Statistics Working Paper 104, Paris, France.

Calmfors, L. and Forslund, A. (1991), "Real Wage Determination and Labour Market Policies: the Swedish Experience", Economic Journal, 101, p.1130-1149.

Calmfors, L. and Driffill, J. (1988), "Centralisation of Wage Bargaining and Macroeconomic Performance", Economic Policy, 4.

Cansier, D. (1993), "Umweltökonomie", UTB für Wissenschaft, Bd. 1749, Stuttgart, Jena, Germany.

Capros, P. and Karadeloglou, P. (1992), "Energy and Carbon Tax: a Quantitative Analysis Using the HERMES/MIDAS Model", in "An Energy Tax in Europe", SEO Conference, Amsterdam.

Capros, P. et al. (1997), "Pre-Kyoto Scenario for the European Union", Final Report to the European Commission, DG-XVII/A2.

Carlevaro, F., Garbely, M. and Müller, T. (1992), "Vers une modélisation en équilibre général des mesures de politique énergétique en Suisse", Serie de Publications du CUEPE No. 49, Université de Genève.

Carraro, C. (1996), "Environmental Fiscal Reform and Unemployment: an Overview" Paper presented at the Unesco-ICEC 1996, Rome, 4-8 March 1996.

Carraro, C. and Soubeyran, A. (1994), "Environmental Policy and the Choice of Technology" in Carraro, C., Katsoulacos, Y. and Xepapadeas, A., op. cit.

Carraro, C. and Siniscalco, D. (1992), "The International Dimension of Environmental Policy", European Economic Review, 36: 379-387.

Carraro, C. and Siniscalco, D. (1993), "Strategies for the International Protection of the Environment", Journal of Public Economics, 52: 309-328.

Carraro, C. and Siniscalco, D. (1994), "Environmental Policy Re-considered: the Role of Technological Innovation", European Economic Review, 38:

Carraro, C. and Siniscalco, D. (1995), "Voluntary Agreements in Environmental Policy: a Theoretical Appraisal", Forthcoming in Xepapadeas, A. (Ed.), *Economic Policy for the Environment and Natural Resources*, E. Elgar, Oxford.

Carraro, C. and Siniscalco, D. (1996), "Environmental Fiscal Reform and Unemployment", Dordrecht, Kluwer Academic Pub.

Carraro, C. and Topa, G. (1995), "Taxation and Environmental Innovation", in Carraro, C. and Filar, J. (Eds.), *Control and Game Theoretic Models of the Environment*, Boston, Birckauser.

Carraro, C. and Galeotti, M. (1994), "WARM: World Assessment of Resource Management. A Technical Report", GRETA, Venice.

Carraro, C. and Galeotti, M. (1995), "Voluntary Agreements on Environmental Innovation: a Macroeconomic Assessment", Paper prepared for the 1995 AFSE Conference, Nantes, 8-9 June 1995.

Carraro, C. and Galeotti, M. (1996a), "Environmental Fiscal Reforms in a Federal Europe", in Braden, J. and Proost, S. (Eds.), *Economic Aspects of Environmental Policymaking in a Federal State*, Edward Elgar, Oxford.

Carraro, C. and Galeotti, M. (1996b), "WARM: A European Model for Energy and Environmental Analysis", Forthcoming in Environmental Modelling and Assessment, 2, 1996; (with Galeotti, M.).

Carraro, C. et al. (1994), "Endogenous Technical Change in Econometric Models of Environmental Policy: Issues and Proposed Strategies", Nota di lavoro 79.94, Fondazione Eni Enrico Mattei (FEEM), Milan, Italy.

Carraro, C., Lanza, A. and Tudini, A. (1994), "Technological Change, Technology Transfers and the Negotiation of International Environmental Agreements", in *International Environmental Affairs*.

Carraro, C., Galeotti, M. and Gallo, M. (1996), "Environmental Taxation and Unemployment: some Evidence on the Double Dividend Hypothesis in Europe", Forthcoming in the Journal of Public Economics.

Carraro, C. and Siniscalco, D. (1992), "Environmental Innovation Policy and International Competition", in *Environmental Resource Economics, 2*.

Carraro, C., Siniscalco, D. (1996), "Voluntary Agreements in Environmental Policy: a Theoretical Appraisal", in Oates, W.E. (Ed.) *Economic Policy for the Environment and Natural Resources: Techniques for the Management and Control of Pollution*, p. 80-94.

Carraro, C., Katsoulacos, Y. and Xepapadeas, A. (Eds.) (1996), "Environmental Policy and Market Structure", Kluwer Academic Pub, Dordrecht.

Carraro, C. (1997), "The Structure of International Agreements on Climate Changes", Paper presented at FEEM-EMF Stanford-IPCC Conference on "International Environmental Agreements on Climate Change", Venice, May 6-7, 1997.

Carraro, C. and Topa, G. (1994), "Should Environmental Innovation Policy Be Internationally Co-ordinated?" in Carraro, C. (Ed.), *Trade, Innovation, Environment*, Kluwer Academic Pub., Dordrecht.

CEC (1993), DGXI, "Un Redéploiment Fiscal au Service de l'Emploi", Brussels, November 1993.

Chao, H. and Peck, S. (1997), "Optimal Environmental Control and Distribution of Cost Burden for Global Climate Change", Journal of International and Comparative Economics, Forthcoming.

Chen, X. (1994), "Substitutions of Information for Energy", Energy Policy, January, p. 15-27.

Cline, W.R. (1992), "The Economics of Global Warming", Washington, D.C, Institute for International Economics, USA.

Clinton, W.J. and Gore, A. (1993), "The Climate Change Action Plan", Washington, DC, USA.

COGGER (1993), "Canadian Options for Greenhouse Gas Emission Reduction", Final Report of the COGGER Panel to the "Canadian Global Change Program and the Canadian Climate Program Board", Canadian Global Change Program Technical Report Series n.93.1, The Royal Society of Canada, September.

Coherence (1991), "Cost-Effectiveness Analysis of CO_2 Reduction Options", CEC DGXII Joule Synthesis Report.

Compte, O. and Jehiel, P. (1997), "International Negotiations and Dispute Resolution Mechanism: the Case of Environmental Negatiations", in Carraro, C. (Ed.), *International Environmental Agreements: Strategic Policy Issues*, Elgar, E., Cheltenham.

Conrad, K. (1993), "Taxes and subsidies for Pollution-Intensive Industries as Trade Policy", in Journal of Environmental Economics and Management 25, p. 121-135.

Conrad, K. and Henseler-Unger, I. (1986), "Applied General Equilibrium Modelling for Long-Term Energy Policy in Germany", Journal of Policy Modelling 8, p. 531-549.

Conrad, K. and Ehrlich, M. (1993), "The Impact of Embodied and Disembodies Technical Progress on Productivity Gaps - An Applied General Equilibrium Analysis for Germany and Spain", Journal of Productivity Analysis 4, p. 317-335.

Conrad, K. and Schmidt, T.F.N. (1996a), "National Economic Impacts of an EU Environmental Policy - An Applied General Equilibrium Analysis", in Proost, S. and Braden, J.B. (Eds.), *Climate Change, Transport and Environmental Policy: Empirical Applications in a Federal System*.

Conrad, K. and Schmidt, T.F.N. (1996b), "Economic Impacts of a Non Co-ordinated vs. a Co-ordinated CO_2 Policy in the EU - An Applied General Equilibrium Analysis", Economic Systems Research, Special Issue, Forthcoming 1998.

Darmstadter, J. et al. (1977), "How Industrial Societies Use Energy", John Hopkins University Press, Baltimore.

Dean, A. and Hoeller, P. (1992), "Costs of Reducing CO_2 Emissions: Evidence from Six Global Models", Organisation for Economic Co-operation and Development, (OECD), Economic Studies, n.19, winter, Paris, France.

Dean, J.M. (1992), "Trade and Environment: A Survey of Literature", in Patrick Low (Ed.), *International Trade and the Environment*, World Bank Discussion Papers, The World Bank, Washington, DC, USA.

Dixon, E.J. et al. (1991), "Afforestation and Forest Management Options and Their Costs at the Site Level", Paper from Technical Workshop to Explore Options for Global Forestry Management, (Eds.) International Institute for Environment and Development, Conference held in April 24-30, 1991, Bangkok, Thailand.

Dixon, E.J. et al. (1991), "Assessment of Promising Forest Management Practices and Technologies for Enhancing the Conservation and Sequestration of Atmospheric Carbon and Their Costs at the Site Level", Report of the US Environmental Protection Agency, Environmental Research Laboratory, Corvallis, OR.

Dixon, E.J. et al. (1994), "Integrated Land-Use Systems: Assessment of Promising Agroforest and Alternative Land-Use Practices to Enhance Carbon Conservation and Sequestration", Climatic Change, 30, p.1-23.

Dobson, A. (1994), "Multifirm Unions and the Incentive to Adopt Pattern Bargaining in Oligopoly", European Economic Review 38, p. 87-100.

Donni, E., Valette, P., Zagamé, P. (1993), "Hermes: Harmonised Econometric Research for Modelling Economic Systems", North-Holland.

Dowlatabadi, H. et al. (1993), "A Model Framework for Integrated Assessment of the Climate Problem", Energy policy, Vol. 21, p. 209-221.

Downing, P. B and White, L. J. (1986), "Innovation in Pollution Control", Journal of Environmental Economics and Management, 13, p. 18-29.

Downing, P.B., White, L.J. (1986), "Innovation in Pollution Control", in Journal of Environmental Economics and Management 13, p. 18-29.

Drèze, J.H., Malinvaud, E. et al. (1993), "Growth and Employment: the Scope of a European Initiative", mimeo, CORE, Louvain.

Dröll, P. (1997), "Negotiated Solutions at Community Level", Paper presented at the Workshop "Negotiated Solutions to Environmental problems" at Utrecht University, March 11th-14th 1997.

Echia, G. and Mariotti, M. (1994), "A Survey on Environmental Policy: Technological Innovation and Strategic Issues", Fondazione Eni Enrico Mattei (FEEM), Working Paper, Milan, Italy.

Edmonds, J and Wise, M. (1997), "Stabilizing the Atmosphere: the Regional Temporal Distribution of Costs under Hypothetical Protocols", Paper presented at FEEM-EMF Stanford-IPCC Conference on "International Environmental Agreements on Climate Change", Venice, May 6-7, 1997.

Edmonds, J., Wise, M. and MacCracken, C. (1994), "Advanced Energy Technologies and Climate Change: an Analysis Using the Global Change Assessment Model (GCAM)", Global Environmental Change Program, Pacific NorthWest Laboratory.

Edmonds, J.A. (1992), "Long Term Modelling of the Links between Economics, Technical Progress, and Environment: Evolution of Approaches and New Trends", Mimeo, Pacific Northwest Laboratory, Washington, DC, USA.

Edmonds, J.A. and Reilly, J. (1983), "A Global Energy-Economic Model of Carbon Dioxide Release", Energy Economics, Vol. 5, n.2, April.

Edmonds, J.A. et al. (1993), "Carbon Coalitions. The Costs and Effectiveness of Energy Agreement to Alter Trajectories of Atmospheric Carbon Dioxide Emissions", Draft, Pacific Northwest Laboratories, Feb., Washington, DC, USA.

Edmonds, J.A. et al. (1994), "Advanced Energy Technologies and Climate Change: An Analysis Using GCAM" Presentation to the Air and Waste Management Meeting, April 6, Tempe, AZ.

EMF (1993), "Reducing Global Carbon Emissions. Costs and Policy Options", Stanford University, Stanford.

Endres, A. (1994) Umweltökonomie, Eine Einführung, Darmstadt, Wissenschaftliche Buchgesellschaft

Endres, A., Querner, I. (1993), Die Ökonomie natürlicher Ressourcen, Darmstadt, Germany.

Endres, A., Schwarze, R. (1994), "Das Zertifikatsmodell vor der Bewährungsprobe" in Kloepfer, M. (Ed.) Studien zum Umweltstaat, Umweltzertifikate und Kompensationslösungen aus ökonomischer und juristischer Sicht, Bonn, p. 137-215.

EPA (1989), "Policy Options for Stabilising Global Climate", Report to Congress, Vol. I, Chapters I-VI, United States Environmental Protection Agency.

Europe Environment, (1997a), "Commission proposes a Common EU-WideExcise Duty System" in Europe Environment, n° 496 - March 25, 1997, p. 4-6.

Europe Environment, (1997b), "Industry furious over Commission's proposals", in Europe Environment, n° 501 - June 10, 1997, p. 1.

European Commission, (1996), "Communication to the Council and European Parliament on Environmental Agreements" COM (96) 561.

European Commission, (1997), "Environmental Taxes and Charges in the Single Market", COM (97) 9.

Ewers, H.-J., Brenck, A. (1995), "Ökonomische Lösungen des Problems der Gefährlichkeit von Stoffen", in Enquete-Kommission Schutz des Menschen und der Umwelt des Deutschen Bundestages (Hrsg.), Umweltverträgliches Stoffstrommanagement, Band 2, Bonn, Germany.

Fisher, B.S., Barrett, S., Bohm, P., Kuroda, M., Mubazi, J.K.E., Shah, A., Stavins, R.N. (1996), "An Economic Assessment of Policy Instruments for Combatting Climate Change", in IPCC (1996), p. 397-439.

Ford Foundation, (1974), "A Time to Choose: America's Energy Future", Energy Policy Project, Cambridge, Mass., Ballinger.

Freeman, R. and Gibbons, R. (1993), "Getting Together and Breaking Apart: the Decline of Centralized Collective Bargaining", Mimeo.

FRN (1987), "Surprising Future", Notes from an International Workshop on "Long-Term World Development", Swedish Council for Planning and Coordination of Research, Report 87:1, Stockolm.

Gal, F. and Frick, P. (1987), "Problem-Oriented Participative Forecasting", Futures, 19(1).

Glachant, M. (1994), "Voluntary Agreements: Bargaining and Efficiency: in Laroui, F. (Ed.) *Environmental Economics*, Amsterdam, SEO-report No. 343, p. 37-54.

Glachant, M. (1996), "The Cost Efficiency of Voluntary Agreements for Regulating Industrial Pollution: A Coasean Approach", Paper presented at the International Workshop on "The Economics and Law of Voluntary Approaches in Environmental Policy, Organised by Fondazione Eni Enrico Mattei and CERNA (Ecole des Mines de Paris), Venice, November 18-19, 1996.

Glomsrod, S. et al. (1990), "Stabilisation of Emissions of CO_2: A CGE Assessment", Cent. Bur. Stat. Discussion Paper n.48, Oslo, Norway.

Goldemberg, J., Squitieri, R., Stiglitz, J., Amano, A., Shaoxiong, X., Saha, R. (1996), "Introduction: Scope of Assessment" in *An Economic Assessment of Policy Instruments for Combatting Climate Change*, in IPCC (1996), p. 17-51.

Goulder, L. (1996), "Optimal Taxation in a Second Best World", Forthcoming in the Yearbook of Environmental Economics.

Goulder, L. H. (1995), "Environmental Taxation and the 'Double Dividend': A Reader's Guide", in *International Tax and Public Finance*, Vol. 2, n°2, p.157-184.

Goulder, L.H. (1995a), "Effects of Carbon Taxes in an Economy with Prior Tax Distortions: An Intertemporal General Equilibrium Analysis", in Journal of Environmental Economics and Management, Vol. 29, No.3, p. 271-297.

Goulder, L.H. (1995b), "Environmental Taxation and the 'Double Dividend': A Reader's Guide", *International Tax and Public Finance*, Vol. 2, p. 157-184.

Grossman, G. (1995), "Pollution and Growth: What do we Know?", in Goldin, I. and Winters, A. (Eds.), *The Economics of Sustainable Development*, Cambridge, Cambridge University Press.

Grubb, M. (1994), "Improved Models", Paper presented at the FEEM-IPCC Conference on "Top-down vs. Bottom-up Models", Milan, 17-20 April 1994.

Grubb, M. (1997), "Technologies, Energy Systems and the Timing of CO_2 Emissions Abatement - An Overview of Economic Issues", Energy Policy, Vol. 25, No. 2, p. 159-72, Elsevier Science Ltd, 1997.

Grubb, M. et al. (1993a), "The Costs of Limiting Fossil-Fuel CO_2 Emissions: A Survey and Analysis", Annual Review of Energy and Environment, Vol. 18, p. 397-478.

Grubb, M. et al. (1993b), "Optimising Climate Change Abatement Responses on Inertia and Induced Technology Development", in *Integrative Assessment of Mitigation. Impacts and Adaptation to Climate Change*, Proceedings of a Workshop held on 13-15

October 1993 at International Institute for Applied Systems Analysis (IIASA), Laxenburg, Austria.

Grubb, M., Edmonds, J. et al. (1993), "The Costs of Limiting Fossil-Fuel CO_2 Emissions", Annual Review of Energy and Environment 18, p. 397-478.

Hall, S., Mabey, N. and Smith, C. (1994), "Macroeconomic Modelling of International Carbon Tax Regimes", Department of Economics Discussion Paper n.94-08, University of Birmingham.

Hansmeyer, K.-H, Schneider, H. K. (1990), Umweltpolitik, ihre Fortentwicklung unter marktsteuernden Aspekten, Göttingen, Germany.

Harrison, W. G. and Rutherford, T. F. (1997), "Burden Sharing, Joint Implementation, and Carbon Coalitions", Paper presented at FEEM-EMF Stanford-IPCC Conference on "International Environmental Agreements on Climate Change", Venice, May 6-7, 1997.

Harrison, W. G., Rutherford, T. F. and Tarr, D. (1997), "Quantifying the Outcome of the Uruguay Round", The Economic Journal, 1997.

Hartkopf, G., Bohne, E. (1983), Umweltpolitik 1: Grundlagen, Analysen und Perspektiven. Opladen, Germany.

Helfland, G.E. (1991), "Standards Versus Standards: The Effect of Different Pollution Restrictions", American Economic Review, 81, p. 622-634.

Heller, T. (1997), "Early Action, Additionality, and the China Trap", Paper presented at FEEM-EMF Stanford-IPCC Conference on "International Environmental Agreements on Climate Change", Venice, May 6-7, 1997.

Hemmelskamp, J. (1997), "Environmental Policy Instruments and their Effects on Innovation" in European Planning Studies, Vol. 5, No. 2, p. 177-194.

Hinchy, M. and Fisher, B. (1997), "Negotiating Greenhouse Abatement and Theory of Public Goods", Paper presented at FEEM-EMF Stanford-IPCC Conference on "International Environmental Agreements on Climate Change", Venice, May 6-7, 1997.

Hoel, M. (1992), "The Role and Design of a Carbon Tax in International Climate Agreement", in Organisation for Economic Co-operation and Development, (OECD) (Ed.) Climate Change. Designing a Practical Tax System, Paris, p. 101-116.

Hoel, M. (1994), "Environmental Policy as a Game between Governments when Plant Locations are Endogenous".

Hoel, M. (1997), "International Co-ordination of Environmental Taxes", in Carraro, C. and Siniscalco, D. (Eds.), New directions in the Economic Theory of the Environment, Cambridge University Press, Cambridge.

Hohmeyer, O. and Koschel, H. (1995), "Umweltpolitische Instrumente zur Förderung des Einsatzes integrierter Umwelttechnik. Gutachten im Auftrag des Büros für Technikfolgen-Abschätzung", Bonn.

Holmlund, B. and Zetterberg, J. (1991), "Insider Effects in Wage Determination", European Economic Review 35, p. 1009-1034.

Horton, G.R. et al. (1992), "The Implications for Trade of Greenhouse Tax Emission Control Policies", Working Paper prepared for the UK Department of Trade and Industry.

Hourcade, J. C. (1997), "Policy Instruments, Timing and Size of Climate Coalitions: Negotiations Patterns and Problems of Revelation of Informations", Paper presented at FEEM-EMF Stanford-IPCC Conference on "International Environmental Agreements on Climate Change", Venice, May 6-7, 1997.

Hourcade, J.C. and Chapuis, T. (1993), "No Regret Potentials and Technical Innovation: A Viability Approach to Integrative Assessment of Climate Policies", *Integrative Assessment of Mitigation. Impacts and Adaptation to Climate Change*, Proceedings of a Workshop held on 13-15 October 1993 at International Institute for Applied Systems Analysis (IIASA), Laxenburg, Austria.

Intergovernmental Panel on Climate Change (IPCC) (1992), "Climate Change: The Supplementary Report to the IPCC Scientific Assessment", Cambridge University Press, Massachusetts.

Intergovernmental Panel on Climate Change (IPCC) (1994), "Energy Supply Mitigation Options", Working Group II, Draft, July 1994.

Intergovernmental Panel on Climate Change (IPCC) (1995), 2nd Assessment Report, 11th Session, Working Group III.

Intergovernmental Panel on Climate Change (IPCC) (1996), "Climate Change 1995", Contribution of Working Group III to the Second Assessment Report of the Intergovernmental Panel on Climate Change, Cambridge, USA.

International Association for Energy Economics (IAEE) (1993), "Energy Environment and Sustainable Development: Challenges for the 21st Century", 16th Annual International Conference of the International Association for Energy Economics, Nusa Dua - Bali, Indonesia, 27-29 July 1993.

International Energy Agency/Organisation for Economic Co-operation and Development (IEA/OECD), (1995), Freiwillige Selbstverpflichtungen der Industrie zur Minderung der CO_2-Emissionen im Energiebereich, Press Release, Paris.

IPSEP (1993), "Energy Policy in the Greenhouse: the Economics of Energy Tax and Non Price Policies", International Project for Sustainable Energy Paths, LBL, California.

Jaffe, A.B., Peterson, S.R., Portney, P.R. and Stavins, R.N. (1995), "Environmental Regulation and the Competitiveness of U.S. Manufacturing: What does the Evidence Tell us?", Journal of Economic Literature, Vol. 33, p. 132-163.

Jensen, P. (1994), "Unemployment and Minimum Wages: A Microeconometric Analysis", Mimeo.

Johansson, T.B. and Swisher, J. (1993), "Perspectives on Bottom-uUp Analysis of Costs of Carbon Dioxide Emission Reductions", Organisation for Economic Co-operation and Development/ International Energy Agency (OECD/IEA), Conference on the "Economics of Global Change", Paris, France, June 1993.

Jorgenson, D.W. and Wilcoxen, P.J. (1994), "Reducing U.S. Carbon Emissions: An Econometric General Equilibirum Assessment", in Gaskins, D. and Weyant, J. (Eds.), *The Cost of Controlling Greenhouse Gas Emissions*, Stanford, CA, Stanford University Press.

Jorgenson, D.W. and Wilcoxen, P.J. (1990), "Intertemporal General Equilibrium Modelling of U.S. Environmental Regulation", Journal of Policy Modeling 12, p. 715-744.

Jung, C., Krutilla, K., Boyd, R. (1996) "Incentives for Advanced Pollution Abatement Technology at the Industry Level: An Evaluation of Policy Alternatives", in Journal of Environmental Economics and Management 30, p. 95-111.

Karp, L. and Newbery, D. (1991), "OPEC and the US Oil Import Tariff", in The Economic Journal, 101, p. 303-313

Karp, L. and Sacheti, S. (1997), "Dynamics and Limited Cooperation in International Environmental Agreements", Paper presented at FEEM-EMF Stanford-IPCC Conference on "International Environmental Agreements on Climate Change", Venice, May 6-7, 1997.

Katz, H. (1993), "The Decentralisation of Collective Bargaining: a Literature Review and a Comparative Analysis", Industrial and Labour Relations Review, 47.

Killingsworth, S. and Heckman, J. (1986), "Female Labor Supply: A Survey", in Ashenfelter, O. and Layard, R. (Eds.), *Handbook of Labor Economics*, Amsterdam, North-Holland.

Klaassen, G. (1996a), "Acid Rain and Environmental Degradation", *The Economics of Emission Trading*, Cheltenham, UK.

Klaassen, G. (1996b), "Emission Trading for Air Quality Standards: Opening Pandora's Box?", Paper prepared for the Annual Meeting of EAERE, June 1996, Lisbon.

Kloepfer, M. (1991), "Zu den neuen umweltrechtlichen Handlungsformen des Staates", in *Juristen Zeitung*, 46. Jg., p. 737-788.

Kohlhaas, M., Praetorius, B. (1994), "Selbstverpflichtungen der Industrie zur CO_2-Reduktion", Berlin.

Kolstad, C. (1987), "Uniformity versus Differentiation in Regulating Externalities", in Journal of Environmental Economics and Management 14, p. 386-399.

Kolstad, C.D. (1993), "Looking Versus Leaping: The Timing of CO_2 Control in the Face of Uncertainty and Learning" in *Costs Impacts, and Benefits of CO_2 Mitigation*, International Institute for Applied Systems Analysis (IIASA), Laxenburg, Austria.

Konishi, H., Le Breton, M. and Weber, S. (1997), "Stable Coalition Structures for the Provision of Public Goods", in Carraro, C. and Siniscalco, D. (Eds.), *New directions in the Economic Theory of the Environment*, Cambridge University Press, Cambridge.

Koschel, H., Brockmann, K.L., Schmidt, T.F.N., Stronzik, M., Bergmann, H. (1997), "Handelbare SO_2-Zertifikate für Europa", Heidelberg (Forthcoming).

Koschel, H., Stronzik, M. (1996), "Impact of Tradable Permits on Innovation", Paper presented at the "Greening of Industry Network" Conference 25-27 November 1996, Heidelberg, Germany.

Koutstaal, P., Nentjes, A. (1995), "Tradable Carbon Permits in Europe: Feasibility and Comparison with Taxes", in Journal of Common Market Studies, Vol. 33, No. 2, p. 219-233.

Kram, T. (1993), "National Energy Options for Reducing CO_2 Emissions", Vol. I, Netherlands Energy Research Foundation, ECN.

Krause, F., Haites,E., Howarth, R. and Koomey, J. (1993), "Cutting Carbon Emissions: Burden or Benefit? The Economics of Energy Tax and Non Price Policies, International Project for Sustainable Energy Paths, El Cerrito, CA., USA.

Laffont, J.J. and Tirole, J. (1994), "Compliance and Innovation Strategies", Paper presented at the 50th IIPF Conference, Harvard, 24-28 August 1994.

Layard, R., Jackman, R. and Nickell, S. (1991), "Unemployment", Oxford, Blackwell.

Lovins, A. (1977), "Soft Energy Paths: Towards a Durable Peace", Harmondsworth, Penguin Books.

Lovins, A. et al. (1981), "Least-Cost Energy-Solving the CO_2 Problem", Andover, Mass., Brick House Publishing.

Manne, A. (1996a), "Hedging Strategies for Global Carbon Dioxide Abatement: A Summary of Poll Results", Draft, Previous Version Presented in August 1995 at the EMF Summer Workshop, "Climate Impacts and Integrated Assessment of Climate Change in Snowmass", Colorado.

Manne, A. (1996b), "Intergenerational Altruism, Discussing and the Greenhouse Debate", Draft, Stanford University.

Manne, A. and Richels, G. (1992), "Buying Greenhouse Insurance", *The Economic Costs of Carbon Dioxide Emission Limits*, Cambridge, MIT Press.

Manne, A. and Richels, G. (1993), "MERGE: A Model for Evaluating Regional and Global Effects of GHG Reduction Policies", Electric Power Research Institute, Palo Alto, CA.

Manne, A. and Oliveira-Martins, J. (1994), "Comparison of Model Structure and Policy Scenarios: GREEN and 12RT", Draft, Annex for the WP1 Paper on "Policy Response to the Threat of Global Warming", Organisation for Economic Co-operation and Development, (OECD), Model Comparison Project II.

Manne, A. and Schrattenholzer. L. (1993), "Global Scenarios for Carbon Dioxide Emissions", Energy, 18 (12), December, p. 1207-22.

Manne, A. and Richels, R. (1997), "On Stabilising CO_2 Concentrations: Cost Effective Emission Reduction Strategies", Paper presented at FEEM-EMF Stanford-IPCC Conference on "International Environmental Agreements on Climate Change", Venice, May 6-7, 1997.

218

Manne, A.S. (1992), "Global 2100: Alternative Scenarios for Reducing Carbon Emissions", Organisation for Economic Co-operation and Development, (OECD), Economics Department Working Papers, n.118.

Manne, A.S. and Richels, R. (1992), "Buying Greenhouse Insurance - The Economic Costs of CO_2 Emission Limits", MIT Press, Cambridge.

Manne, A.S. and Olsen, T.R. (1994), "Greenhouse Gas Abatement - Toward Pareto-Optimal Decisions under Uncertainty", Department of Economics Discussion Paper n.94-06, University of Birmingham.

Martin, J.M. (1992), "Economie et politique de l'énergie", Paris, Armand Collin, coll. Cursus Economie, 192.

Masera, O. et al. (1994), "Forest Management Options for Sequestering Carbonin Mexico", Biomass and Bioenergy, Forthcoming.

Mayeres, I. and Proost, S. (1997), "Optimal Tax and Public Investment Rules for Congestion Type of Externalities", in Scandinavian Journal of Economics 99(2), p. 261-279.

McCann, R.J. and Moss, S.J. (1993), "Nuts and Bolts: The Implications of Choosing Greenhouse-Gas Emission Reduction Strategies", Policy Study n.171, Reason Foundation, Los Angeles, California.

McDonald, I.M. and Solow, R.M (1981), "Wage Bargaining and Employment", American Economic Review, 71, p. 896-908.

McKibbin, W.J. and Wilcoxen, P.J. (1992a), "The Global Costs of Policies to Reduce Greenhouse Gas Emission", Brookings Discussion Paper in International Economics n.97, The Brookings Institution, Washington.

McKibbin, W.J. and Wilcoxen, (1992b), "G. Cubed: A Dynamic Multi-Sector General Equilibrium Model of the Costs of the Global Economy (Quantifying the Costs of Curbing CO_2 Emission)", Brookings Discussion Paper in International Economics n.98, The Brookings Institution, Washington.

McKibbin, W.J. and Wilcoxen, P.J. (1997), "A Better Way to Slow Global Climate Change", Australian National University, March.

Meadows, D. (1972), "The Limits to Growth", London.

Metz, B. (1997), "Voluntary Approaches: Some Results from IEA Experiences", Paper presented at FEEM-EMF Stanford-IPCC Conference on "International Environmental Agreements on Climate Change", Venice, May 6-7, 1997.

Milliman, S. R. and Prince, R. (1989), "Firm Incentives to Promote Technological Change in Pollution Control", Journal of Environmental Economics and Management 17, p. 247-265.

Mintzer, I. (1988), "Living in a Warmer World: Challenges for Policy Analysis and Management", Journal of Policy Analysis and Management, Vol. 7, n.3, p. 445-59.

Moene, K. (1998), "Unions' Threats and Wage Determination", Economic Journal, 96, p. 98-109.

Mongia, N. et al. (1991), "Cost of Reducing CO_2 Emissions from India", Energy Policy, 19 (10), p. 978-987.

Montgomery, W. D., Bernstein P. M. and Rutherford, T. (1997), "Does Trade Matter? Model Specification and Impacts on Developing Countries", Paper presented at FEEM-EMF Stanford-IPCC Conference on "International Environmental Agreements on Climate Change", Venice, May 6-7, 1997.

Montgomery, W.D. (1972), "Markets in Licenses and Efficient Pollution Control Programs", in Journal of Economic Theory, Vol. 5, p 395-418.

Moro, A. (1993), "A Survey on R&D and Technological Innovation: Firms' Behaviour, Regulation and Pollution Control", Fondazione Eni Enrico Mattei (FEEM), Working Paper n.73.93, Milan, Italy.

Morrison, C.J. (1988), "Quasi-Fixed Inputs in U.S. and Japanese Manufacturing: A Generalized Leontief Restricted Cost Function Approach", Review of Economics and Statistics LXX, p. 75-287.

Moulton, R. and Richards, K. (1990), "Costs of Sequestering Carbon Through Tree Planting and Forest Management in the US", General Technical Report WO-58, US Department of Agriculture, Washington, DC, USA.

Muller, F. (1996), "Energy Taxes, the Climate Change Convention, and Economic Competitiveness", in Hohmeyer, O., Otinnger, R.L. and Rennings, K. (Eds.), *Social Costs and Sustainability*, p. 465-487.

Munasinghe, M. and Munasinghe, S. (1993), "Barriers to and Opportunities for Technological Change in developing Countries to Reduce Global Warming", Paper presented at the Montreal Meting of the IPCC Working Group III, Montreal, 3-7 May, 1993.

Murdoch, J., Sandler, T. and Sargent, K. (1997), "A Tale of Two Collectives: Sulfur Versus Nitrogen Oxides Emission Reduction in Europe", Forthcoming, Economica.

Murswiek, D. (1988), "Freiheit und Freiwilligkeit im Umweltreecht", in Juristen Zeitung, 43. Jg., Heft 21, p. 985-993.

Musu, I. (1994), "On Endogenous Sustainable Growth", Working Paper, University of Venice.

National Academy of Sciences (1992), "Policy Implications of Greenhouse Warming: Mitigation, Adaptation, and the Science Base", National Academy Press, Washington, DC, USA.

New York State (1991), "Analysis of Carbon Reduction in New York State", Report of the New York State Energy Office, NYS Energy Office, New York.

Nordhaus, W.D. (1977), "Strategies of the Control of Carbon Dioxide", Cowles Foundation Discussion Paper, n.443, Yale University, New Haven, Conn.

Nordhaus, W.D. (1979), "The Efficiency Use of Energy Resources", New Haven, Conn, Yale University Press.

Nordhaus, W.D. (1991a)," To Slow or Not to Slow The Economic of the Greenhouse Effect", The Economic Journal, Vol. 101, p. 920-937.

Nordhaus, W.D. (1991b), "The Cost of Slowing Climate Change: A Survey", The Energy Journal, 12(1), p. 37-65.

Nordhaus, W.D. (1994), "Managing the Global Commons: The Economics of Climate Change", MIT Press, Forthcoming.

Nordhaus, W.D. and Yang, Z. (1996), "A Regional Dynamic General-Equilibrium Model of Alternative Climate-Change Strategies", The American Economic Review, September 1996.

Oates, W. (1991), "Pollution Charges as a Source of Public Revenues", Resources for the Future, Discussion Paper QE92-05, Washington DC, USA.

Organisation for Economic Co-operation and Development (OECD), (1994), "Managing the Environment. The Role of Economic Instruments", Paris, France.

Organisation for Economic Co-operation and Development (OECD), (1994), "The Costs of Cutting Carbon Emissions: Results from Global Models", Paris.

Organisation for Economic Co-operation and Development (OECD) (1997), "Evaluating Economic Instruments for Environmental Policy", Paris.

Official Journal of European Commission, (1992), N° C 196/1 of 3 August 1992.

Oliveira-Martins, J. et al. (1992), "The Costs of Reducing CO_2 Emissions: A Comparison of Carbon Tax Curves with GREEN", Organisation for Economic Co-operation and Development (OECD), Economics Department Working Papers, n.118.

Olivier, D. et al. (1982), "Energy-Efficient Futures: Opening the Solar Option", London, Earth Resources Research.

Parks, P. and Hardie, I. (1995), "Least Cost Forest Carbon Reserves: Cost-Effective Subsidies to Convert Marginal Agricultural Land to Forest", Forthcoming in Land Economics.

Parry, I. (1994), "Pollution Taxes and Revenue Recycling", Working Paper, U.S. Department of Agriculture.

Peck, S. and Teisberg, T.J. (1993), "Global Warming Uncertainties and the Value of Information: An Analisys Using CETA", Resource and Energyeconomics, 15, p. 71-97.

Peck, S. and Teisberg, T.J. (1997a), "International CO_2 Emissions Targets and Timetables: An Analysis of the AOSIS Proposal", Environmental Modeling and Assessment, Vol. 1, No.4, February.

Peck, S. and Teisberg, T.J. (1997b), "CO_2 Concentration Limits, The Cost and Control, and The Potential for International Agreement", Paper presented at FEEM-EMF Stanford-IPCC Conference on "International Environmental Agreements on Climate Change", Venice, May 6-7, 1997.

Pencavel, J. (1991), "Labour Markets under Trade Unionism", Oxford, Blackwell.

Pethig, R. (1996), "Ecological Tax Reformand Efficiency of Taxation: A Public Good Perspective", (Revised Version, July 1996), Diskussionbeitrag Nr. 57-96, Universität Siegen.

Pezzey, J. (1992), "Analysis of Unilateral CO_2 Control in the Europen Community and OECD", The Energy Journal, 13, p. 159-71.

Proost, S. and Van Regermorter, D. (1991), "Economic Effects of a Carbon Tax. With a General Equilibrium Illustration for Belgium", Public Economy Research Paper n.11, Katholieke Univ.Leuven, Belgium.

Rauscher, M. (1995), "Environmental Regulation and the Location of Polluting Industries", in International Tax and Public Finance 2, p. 229-244.

Ravindranath, N. and Somashekhar, B. (1994), "Potential and Economics of Forestry Options for Carbon Sequestration in India", Biomass and Bioenergy, Forthcoming.

Rehbinder, E. (1994), "Übertragbare Emissionsrechte aus juristischer Sicht, Teil I, II und III", in Michael Kloepfer (Ed.) Studien zum Umweltstaat, Umweltzertifikate und Kompensationslösungen aus ökonomischer und juristischer Sicht. Bonn, p. 28-136 and p. 216-255.

Reichmann, H. (1994), Umweltabgaben. Theoretische Grundlagen, Klassifikationen und potentielle Wirkungsbrüche. Frankfurt am Main, Germany.

Reinert, K.A., Roland-Holst, D.W. and David, W. (1992), "Armington Elasticities for United States Manufactoring Sectors", Journal of Policy Modeling, Vol. 14, No. 5, p. 631-639.

Rennings, K., Brockmann, K.L., Bergmann, H. (1997), "Voluntary Agreements in Environmental Protection - Experiences in Germany and Future Perspectives", ZEW-Discussion Paper No. 97-04 E. Mannheim, Germany.

Repetto, R., Dower, R., Jenkins, R. and Geoghegan, J. (1992), "Green Fees: How a Tax Shift Can Work for the Environment and the Economy", World Resource Institute, New York.

Requate, T. (1995), "Incentives to Adopt New Technologies under Different Pollution-Control Policies", in International Tax and Public Finance, 2, p. 295-317.

Richards, K. et al. (1993), "Costs of Creating Carbon Sinks in the US", Chapter in Proceedings of the "International Energy Agency Carbon Dioxide Disposal Symposium", Riemer, P. (Ed.), Pergamon Press, Oxford.

Richels, R. and Edmonds, J. (1993), "The Costs of Stabilising Atmospheric CO_2 Conentrations", in Integrative Assessment of Mitigation. Impacts and Adaptation to Climate Change, Proceedings of a Workshop held on 13-15 October 1993 at International Institute for Applied Systems Analysis (IIASA), Laxenburg, Austria, Forthcoming in Energy Policy.

Richels, R. et al. (1996)," The Berlin Mandate The Design of Cost-Effective Mitigation Strategies", Draft, Paper of the Subgroup on the Regional Distribution of the Costs and Benefits of Climate Change Policy Proposals, EMF-14, Stanford University.

Rose, A., Stevens, B. (1996), "Equity Aspects of the Marketable Permits Approach to Global Warming".

Ross, M. and Williams, P. (1981), "Our Energy: Regaining Control", New York, McGraw Hill.

Rotmans, J. (1990), "IMAGE: An Integrated Model to Assess the Greenhouse Effect", Kluwer Academic Publishers, Dordrecht, Netherlands.

Rotmans, J. et al. (1995), "Global Change and Sustainable Development: A Modeling Perspective for the Next Decade", National Institute of Public Health and Environmental Protection, Bilthoven, Netherlands, June.

Rowe, M.D. and Hill, D. (Eds.) (1989), "Estimating National Costs of Controlling Emissions from the Energy System", Report of the Energy Technology Systems Analysis Project, International Energy Agency (IEA), Brookhaven National Laboratory, New York.

Sartzetakis, E. S. (1997), "Tradeable Emission Permits Regulations in the Presence of Imperfectly Competitive Products Markets: Welfare Implications" in *Environmental and Resource Economics* 9, p. 65-91.

Schmelzer, D. (1996), "Voluntary Agreements in Environmental Policy: Negotiating Emission Reductions", Paper presented at the International Workshop on "The Economics and Law of Voluntary Approaches in Environmental Policy", Organised by Fondazione Eni Enrico Mattei and CERNA (Ecole des Mines de Paris), Venice, November 18-19, 1996

Sedjo, R. and Solomon, A. (1989), "Greenhouse Warming: Abatement and Adaptation", RFF Proceedings, Crosson, P. et al. (Eds.), July, 1989, p.110-119.

Segerson, K. (1995), "Issues in the Choice of Environmental Policy Instruments", in *Environmental Policy with Political and Economic Integration - The European Union and the United States*, Braden, J., Folmer, H., Olen, T. (Eds.), New Horizons in Environmental Economics.

Selvanathan, E. and Selvanathan, S. (1994), "The Demand for Transport and Communications in the UK and Australia", Transportation Research, Vol. 28b, n.1, p.1-9.

Shackleton, R. et al. (1993), "The Efficiency Value of Carbon Tax Revenues", PRISTE-CNRS, October 1992, Paris.

Shukla, P. R. (1997), "Socio-Economic Dynamics of Developing Countries: Some Ignored Dimensions in Integrated Assessment", proceedings of IPCC Symposium on "Integrated Assessment Process", Toulouse, 24-26 October.

Shukla, P. R. (1997), "Implications of Proposed Emissions Limitations for Developing (Non Annex I) Countries", Paper presented at FEEM-EMF Stanford-IPCC Conference on "International Environmental Agreements on Climate Change", Venice, May 6-7, 1997.

Siebert, H. (1992), "Economics of the Environment. Third", Revised and Enlarged Edition, Berlin.

SRU, (Ed.) (1994), "Umweltgutachten 1994 des Rates von Sachverständigen für Umweltfragen", Für eine dauerhaft-umweltgerechte Entwicklung, Drucksache 12/6995, Wiesbaden, Germany.

Stavins, R. (1996), "Correlated Uncertainty and Policy Instruments Choice", in Journal of Environmental Economics and Management 30, p. 218-232

Stern, R.M., Francis, J. and Schumacher, B. (1976), "Price Elasticities in International Trade", Macmillan Press, London.

Stockholm Environment Institute (1993), "Towards s Fossil Free Energy Future. A Technical Analysis for Greenpeace International", SEI, Boston.

Tang, P., de Mooij, R. and Nahuis, R. (1997), "An Economic Evaluation of Alternative Approaches for Limiting the Costs of Unilateral Action to Slow Down Global Climate Change- Simulations with WORLDSCAN", Study Prepared for the EC, CPB Netherlands Bureau for Economic Policy Analysis.

Tietenberg, T.H. (1994), "Market-Based Mechanisms for Controlling Pollution - Lessons from the U.S", in Sterner, T. (Ed.) Economic Policies for Sustainable Development, Dordrecht, Netherlands, p. 20-45.

Tietenberg, T. H. (1983), "Market Approaches to Environmental Protection", in Giersch, H. (Ed.) Reassessing the Role of Government in the Mixed Economy, Symposium 1982, Tübingen, Germany, p. 233-263.

Tietenberg, T.H. (1990), "Economic Instruments for Environmental Regulation", in Oxford Review of Economic Policy. Vol. 6. No. 1, p. 17-33.

Tietenberg, T.H. (1995), "Tradeable Permits for Pollution Control when Emission Location Matters: What have We Learned?", in Environmental and Resource Economics 5, p. 95-113.

Tobey, J.A. (1992), "Economic Issues in Global Environmental Change", Global Environmental Change, September, p. 215-228.

Tol, R. (1997), "The Optimal Timing of Greenhouse Gas Emission Abatement, Individual Rationality and Intergenerational Equity", Paper presented at FEEM-EMF Stanford-IPCC Conference on "International Environmental Agreements on Climate Change", Venice, May 6-7, 1997.

UNEP, (1994), UNEP Greenhouse Gas Abatement Costing Studies, UNEP Collaborating Centre on Energy and Environment, RISØ National Laboratory, Denmark, May 1994.

US Congress, Office of Technology Assessment (1991), "Changing by Degrees: Steps to Reduce Greenhouse Gases", Office of Technology Assessment (OTA), Washington, DC, USA.

USEPA (1993), "Options for Reducing Methane Emissions Internationally", Report to Congress, Hogan, K.B. (Ed.), Office of Air and Radiation, Washington, DC, USA.

USEPA (1994), "Opportunities to Reduce Anthropogenic Methane Emissions in the US", Report to Congress, Hogan, K.B. (Ed.), Office of Air and Radiation, Washington, DC, USA.

Van Kooten, G. et al. (1992), "Potential to Sequester Carbon in Canadian Forests: Some Economic Considerations", Canadian Public Policy, XVII (2), p.127-138.

Vouyoukas, L. (1993), "IEA Medium Term Energy Model", in *The Costs of Cutting Carbon Emissions: Results from Global Models*, Organisation for Economic Co-operation and Development, (OECD), Paris, France.

Wack, P. (1985a), "Scenarios: Uncharted Waters Ahead", Harvard Business Review, n.5.

Wack, P. (1985b), "Scenarios: Shooting the Rapids", Harvard Business Review, n.6.

Wegner, G. (1994), Marktkonforme Umweltpolitik zwischen Dezisionismus und Selbststeuerung, Tübingen, Germany.

Weitzmann, M.L. (1974), "Prices versus Quantities", in Review of Economic Studies 41, p. 477-491.

Whalley, J. and Wigle, R. (1992), "Results for the OECD Comparative Modelling Exercise from the Whalley-Wigle Model", Organisation for Economic Co-operation and Development, (OECD), Economics Department Working Papers, n.121, Paris, France.

Wigley, T. et al. (1995), "Stabilising CO_2 Concentrations: Choosing the Right Emissions Pathway", submitted for publication.

Wigley, T., Richels, R. and Edmonds, J. (1996), "Economic and Environmental Choices in the Stabilization of Atmospheric CO_2 Concentrations", Nature, Vol. 379, 18 January.

Wilson, D. and Swisher, J. (1993), "Exploring the Gap: Top-down Versus Bottom-up Analyses of the Cost of Mitigating Global Warming", Energy policy, March, p. 249-263

Wirl, F. (1994), "Pigouvian Taxation of Energy for Flow and Stock Externalities and Strategic, Non Competitive Energy Pricing", in Journal of Environmental Economics and Management 26, p. 1-18.

Xu, D. (1994), "The Potential for Reducing Atmospheric Carbon by Large-Scale Afforestation in China and Related Cost-Benefit Analysis", Biomass and Bioenergy, Forthcoming.

Yang, Z. (1997), "Necessary Conditions for Stabilization Agreements", Paper presented at FEEM-EMF Stanford-IPCC Conference on "International Environmental Agreements on Climate Change", Venice, May 6-7, 1997.

Yang, Z., Eckaus, R. S, Ellerman, A.D. and Jacoby, H.D. (1996), "The MIT Emissions Prediction and Policy Analysis (EPPA) Model", Working Paper, MIT, 1996.